Human Factors
in Product Design

Human Factors in Product Design:
Current Practice and Future Trends

Edited by
W.S.GREEN and P.W.JORDAN

CRC Press
Taylor & Francis Group
Boca Raton London New York

CRC Press is an imprint of the
Taylor & Francis Group, an **informa** business
A TAYLOR & FRANCIS BOOK

CRC Press
Taylor & Francis Group
6000 Broken Sound Parkway NW, Suite 300
Boca Raton, FL 33487-2742

First issued in paperback 2019

© 1999 by Bill Green and Pat Jordan
CRC Press is an imprint of Taylor & Francis Group, an Informa business

No claim to original U.S. Government works

ISBN-13: 978-0-7484-0829-0 (hbk)
ISBN-13: 978-0-367-39950-4 (pbk)
Library of Congress catalog number: 99-13215

Library of Congress Cataloging-in-Publication Data

Catalog record is available from the Library of Congress

**Visit the Taylor & Francis Web site at
http://www.taylorandfrancis.com**

**and the CRC Press Web site at
http://www.crcpress.com**

ANALYSING ARCHITECTURE NOTEBOOKS

A rchitecture is such a rich and subtle field of human creativity that it is impossible to encapsulate it completely in a single book. I tried to describe some of the basics in an earlier book, *Analysing Architecture*, which has now appeared in four editions, increasing in size each time. But even though that book has almost doubled in content, there is more to cover. So, rather than make the original even heavier, I have decided to add further chapters as a series of separate smaller volumes.

These *Analysing Architecture Notebooks* are the new chapters I would have added to *Analysing Architecture* had not excessive size become a concern. The series format also allows me to explore topics at greater length than if I were confined to just a few extra pages in the original book. Nevertheless, the shared aim remains the same: to explore and expose the workings of architecture in ways that might help those who face the challenges of doing it.

Simon Unwin is Emeritus Professor of Architecture at the University of Dundee in Scotland. Although retired, he continues to teach at the Welsh School of Architecture in Cardiff University, Wales, where he taught for many years. His books are used in schools of architecture around the world and have been translated into various languages.

Books by Simon Unwin

Analysing Architecture
An Architecture Notebook: Wall
Doorway
Exercises in Architecture – Learning to Think as an Architect
Twenty-Five Buildings Every Architect Should Understand
The Ten Most Influential Buildings in History: Architecture's Archetypes

ebooks (available from Apple Books)

Skara Brae
The Entrance Notebook
Villa Le Lac
The Time Notebook

The Analysing Architecture Notebook Series

Metaphor
Curve
Children as Place-Makers
Shadow

Simon Unwin's website is at *simonunwin.com*

Some of Simon Unwin's personal notebooks, used in researching and preparing this and his other books, are available for free download from his website.

Part of a review of *Children as Place-Makers*:

'Through brief one-page meditations, accompanied by Unwin's winsome drawings, on commonplace childhood activities like sand castle building and play forts built of tree branches, he gradually demonstrates how elemental place-making evolves into adult architectural designs. Given the rigorous professional training architects need to attain in order to practice, Unwin's focus might seem like a discounting of the technical skills architects as adults need to master. But in no way do I find his approach anti-intellectual or counter-professional. He is insisting on the root-impulse that drives one to become an architect in the first place, a primal impulse that can all too easily become forgotten during the rigours of academic training and daily professional practice. This book is really a primer about architectural imagination that begins intuitively, unselfconsciously in childhood. That was perhaps what Picasso had in mind when he remarked: "It takes a very long time to become young". Unwin's extraordinary book may reduce the time needed to design with the freshness reflecting a child's vision of place.'

Norman Weinstein – 'Who Isn't a Born Architect?', ArchNewsNow.com, 2019, available at: archnewsnow.com/features/Feature585.htm (September 2019).

ANALYSING ARCHITECTURE NOTEBOOKS

SHADOW

the architectural power of
withholding light

First published 2020
by Routledge
2 Park Square, Milton Park, Abingdon, Oxon OX14 4RN

and by Routledge
52 Vanderbilt Avenue, New York, NY 10017

Routledge is an imprint of the Taylor & Francis Group, an informa business

Publisher's Note
This book has been prepared from camera-ready copy provided by the author.

British Library Cataloguing-in-Publication Data
A catalogue record for this book is available from the British Library

Library of Congress Cataloging-in-Publication Data
Names: Unwin, Simon, 1952- author.
Title: Shadow : the architectural power of withholding light/ Simon Unwin.
Description: Abingdon, Oxon ; New York : Routledge, 2020. | Series: Analysing architecture notebooks | Includes bibliographical references and index.
Identifiers: LCCN 2019056494 (print) | LCCN 2019056495 (ebook) | ISBN 9780367442569 (hardback) | ISBN 9780367442583 (paperback) | ISBN 9781003008583 (ebook)
Subjects: LCSH: Architecture--Composition, proportion, etc. | Shades and shadows in architecture.
Classification: LCC NA2760 .U59 2020 (print) | LCC NA2760 (ebook) | DDC 720--dc23
LC record available at https://lccn.loc.gov/2019056494
LC ebook record available at https://lccn.loc.gov/2019056495

ISBN: 978-0-367-44256-9 (hbk)
ISBN: 978-0-367-44258-3 (pbk)
ISBN: 978-1-003-00858-3 (ebk)

Typeset in Arial and Georgia

by Simon Unwin

for students of architecture

With apologies to Robert Fludd – 'Et sic in infinitum', in *Utriusque cosmi maioris scilicet et minoris metaphysica, physica atqve technica historia*, 1617. See: archive.org/details/utriusquecosmima01flud/page/n33 (November 2019).

'Light is the expeller of darkness. Shadow is the suppression of light.'

Leonardo da Vinci (15th century CE), trans. MacCurdy, 1939.

'The truth is, Darkness will be found in every part of Space, where there is no Light. Should it be asked, What becomes of Darkness, when Light enters into it? We reply, that Darkness doth not make its escape. Darkness appears to be an unlimited negative.'

Joseph Unwin – *Materialism Refuted*, 1829.

CONTENTS

'L'architecture est le jeu savant, correct et magnifique des volumes assemblés sous la lumière. Nos yeux sont faits pour voir les formes sous la lumière; les ombres et les clairs révèlent les formes.'

('Architecture is the masterly, correct and magnificent play of mass brought together in light. Our eyes are made to see forms in light; light and shade reveal these forms.')

Le Corbusier – *Vers Une Architecture* , 1923
(trans. Etchells –*Towards a New Architecture*, 1927.)

'Only
There is shadow under this red rock,
(Come in under the shadow of this red rock),
And I will show you something different from either
Your shadow at morning striding behind you
Or your shadow at evening rising to meet you...'

T.S. Eliot – *The Waste Land*, 1922.

'As we came in the door an elderly waitress... was kneeling by a candle behind which stood a large screen. On the far side of the screen, at the edge of the little circle of light, the darkness seemed to fall from the ceiling, lofty, intense, monolithic, the fragile light of the candle unable to pierce its thickness, turned back as from a black wall.

Junichiro Tanizaki, trans. Harper and Seidensticker (1954) – *In Praise of Shadows* (1933), 2001.

'They were dark caves. Even when they open towards the sun, very little light penetrates down the entrance tunnel into the circular chamber. There is little to see, and no eye to see it, until the visitor arrives for his five minutes, and strikes a match. Immediately another flame rises in the depths of the rock and moves towards the surface like an imprisoned spirit.'

E.M. Forster – *A Passage to India* (1924), 1989.

'Our work is of shadow.'

Louis Kahn (1971), quoted in Alexandra Tyng – *Beginnings: Louis I. Kahn's Philosophy of Architecture*, 1984.

PREFACE

Shadow is a common metaphor in literature, psychology, politics... But this Notebook is about actual shadow and its roles as an element of architecture.

Metaphorical shadows are often if not always cast as negative. Since ancient times the prospect of death has cast its 'shadow' across our lives. Thieves and spies hide 'in the shadows' metaphorically as well as literally. Attention-grabbers like to 'hog the limelight' and put others 'in the shade'. The trope 'a shadow fell across her face' suggests not an actual shadow (though in a movie that might happen) but a flicker of negative emotion – anger, fear, sadness, regret... Jungian psychologists say we all carry with us our own 'shadow' – the aspects (often perceived as negative) of our personalities that we hide away from ourselves as well as others.

We make sense of our world in terms of spectrums with binary extremes: positive—negative; good—bad; right—left; white—black; up—down; forward—backward; male—female; life—death; order—chaos; reliability—fickleness; love—hate; collaboration—competition; day—night; peace—war... Poets, politicians, bigots... find or try to assert correspondences (such is the stuff, and treachery, of metaphor) between the extremes: 'white is good—black is bad'; 'male is reliable—female is fickle'; 'female is for peace and collaboration—male is for competition and war'... Examples are legion. And often the underlying sentiment suggests that while one end of the perceived spectrum holds moral virtue, the other is morally suspect.

In Arnold Böcklin's 'Die Toteninsel' ('Isle of the Dead'; 1886 version) a soul approaches the realm of death, the massed cypress trees offering nothing more of what is to be found there than a portal of the deepest shadow.

Light—shadow is a binary spectrum with underlying moral connotations. It is associated, metaphorically, with some of those listed above. Light is good – revealing, rational, open, happy... Shadow is bad – concealing, superstitious, closed in, sad... Shadow is perceived as light's negative, and as such its suspect, sinister opposite... even its enemy.

Moral connotations associated with metaphorical shadows can affect our attitudes to actual shadows. Yes, there are practical reasons why shadows can be problematic: it is difficult to work if you cannot see what you are doing; shadows provide hiding places for miscreants; the amenity of a house or garden is diminished if it is overshadowed. These problems contribute of course to the employment of shadow as a negative metaphor. But that contribution is reciprocated when we seek to obviate shadows: condemning them generally as 'bad' because of their metaphorical moral connotations; or deepen them because they provide cover for behaviour that we do not want to be seen too clearly (if at all). Law courts and government buildings are brightly lit because light is a metaphor for openness and scrutiny; night clubs are dark because shadow provides cover for bad dancing and intimacy; open-plan offices are evenly lit for practical reasons but also to imply egalitarianism; meditation spaces are dim to lessen distraction and impel our attention inwards.

Shadows have a metaphorical role in the poetry of architecture just as much as in the poetry of words. After all, the narrative of Böcklin's painting (sketched above) depends on the 'Isle of the Dead' being an identified place – a work of architecture – with the 'deep shadow' of death at its conceptual and physical core. But shadow is a practical and aesthetic element of architecture too. This Notebook is about what we, as architects, can do with actual shadows. And, in the main, it explores how shadows make a positive contribution to architecture.

Le Corbusier famously described architecture as a composition of forms 'brought together in light'; (see the top quotation on page viii). His statement can be interpreted as a musical metaphor: architects compose forms like composers compose music; the purpose of both is aesthetic before it is functional. With flutes and violins we play music; with architecture Le Corbusier suggests we orchestrate light. But in his second sentence – 'light *and shade...*' (in Etchell's translation reversed from Le Corbusier's '*les ombres* et les clairs') '... reveal these forms' – he implicitly acknowledges that the primary contribution of architectural forms in daytime is actually to obstruct and modify the sun's light – i.e. to create and modify shadows. As Louis Kahn later said, 'Our work is of shadow' (see the bottom quotation on page viii), a statement that could be interpreted as metaphorical but will here be explored literally.

All art occurs in ambient conditions that it occupies, overlays, usurps, interacts with... For music the ambient condition is silence, represented by that moment of quiet just before the orchestra starts... (Except that some composers, John Cage for example, have experimented with the contribution ambient noise can make to music; e.g. Cage's piece '4'33"' consists of nothing but ambient sounds.) For drawing the ambient condition is usually white paper; though sometimes the chosen ground is black or coloured and different tonal grades of media are used. (The drawn equivalent of Cage's '4'33"' would be a blank, but maybe slightly creased or marked, piece of white paper: maybe Piero Manzoni's 'Achrome', 1958; or Robert Ryman's 'Ledger', 1982.) There are many dimensions to the ambient conditions of architecture – it exists in, contributes to and changes the full complexity of the real world – but a major one is light. For architecture in daytime the ambient condition is light from the sun and sky; at night it is darkness, and artificial light is needed to obviate that darkness and to cast shadows within shadow. Shadow is what I have (in *Analysing Architecture*) termed a 'modifying element' of architecture. Consequent on light that brings it into being, shadow modifies the places we make, contributing to their identity. This might happen by accident – without conscious intent – but the modifying role of shadow in the identification of place can also be determined by you the architect. Shadow modifies but it may also be modified.

For good or ill, a primary result of architecture is to cause shadow. Shadow is an element of architecture. Light may be the ambient blanket under which architecture is created but, as Le Corbusier implies, the play of shadow is its voice and music, its delineation and modelling. By means of architecture we can engineer complete darkness or we can modulate shade with obstructions, openings, filters, artificial illumination and surfaces of varying reflectance. As in the yin-yang symbol (page viii) light and shadow are reciprocal, mutually interactive and interpenetrating. Architecture mediates: it is container, content and interface between the two. It is an instrument for playing shadow.

There are different kinds of shadow. Mostly we think in terms of shadows cast... shadows cast by ourselves or by objects, caused by obstructing a source of light and projected onto a surface – the ground, a wall or screen... In this sense things *have* shadows. But architecture *contains* shadows. Yes, buildings can cast shadows as objects but the surfaces of walls contain the shadows of their incisions and mouldings and, even more powerfully, the rooms within buildings have shadows as ingredients of their atmospheres. And atmospheres vary during the day, in different weather conditions and different parts of the world.

Architectural interiors are places of shadow: shadows modulated and shadows of immersion. Far from being negative, such shadows can be ingredients of the practical and aesthetic – as well as the metaphorical – power of architecture. In hot sunny places shadows help provide comfort (and reduce energy demand). Shadows provide psychological refuge. Shadow-play – of sun projecting the moving shadows of a tree on a wall; of a 'searchlight' sunbeam tracking through a dark room; the gradations of shade in a Japanese tea room... – contributes to our aesthetic appreciation of architecture.

It is too much of a generalisation to suggest that twentieth-century Modernism despised shadow but there are some strands that avoided it. In painting for example, look at the abstraction of Picasso's 'La Rêve', 1932; or the simplified colours of Patrick Caulfield's 'Oh Helen, I Roam My Room', 1973. Modern music can be termed 'atonal' for its own reasons but it is so also because it might – as in the characteristic work of Arnold Schoenberg, Karlheinz Stockhausen or Philip Glass – lack the 'light and shade' of pieces from the eras of Bach, Mozart and Beethoven. In Modern architecture – perhaps to absolve the deep metaphorical shadows of two World Wars or perhaps to manifest an abiding desire for egalitarianism – shadow is often dispelled by large areas of glazing and bright even lighting. Modern architecture's flat planar surfaces are characteristically conceived without intrinsic shadow.

Such banishment of a significant and useful element of architecture seems too harsh, even dogmatic. Perhaps shadow has served its 'punishment' for its association with all things negative. Maybe it is time to explore again its potential in architecture.

This Notebook, as are the others in this series, is intended not merely to describe and illustrate a subject – in this instance the practical, poetic and aesthetic potential of shadow in architecture – but to introduce ideas that you might explore and develop in your own design work, and stimulate your own search for more. We are all prone to an abiding passivity, and enjoy others telling us stories about the world, making sense of it for us and recounting interesting observations. Perhaps we see knowledge as a commodity provided by others, which we might assimilate, or forget. (Such is sometimes seen as the principle currency of education and examinations.) Multiple screen-based media provide copious entertainment that we click on, share and forget.

But for us architects, architecture is not merely a matter of acquiring knowledge provided by others. Nor is it merely scrolling through seductive images on Dezeen or ArchDaily. It is an intellectually exploratory and inventive activity: one we are professionally committed to engage in. To do architecture we need ideas; ideas are our main stock in trade. And ideas come from searching.

There are the beginnings of many ideas about shadow in this Notebook. It is not enough just to read about and them and look at the drawings. You should, as an architect, make deliberate efforts to extend, develop and add to that (limited) stock of ideas by exercising your own curiosity and kicking into gear your capacity for design.

So, become conscious of the role of shadow in the world around you. Look at how other architects have generated and modified them for their own purposes – practical, aesthetic and metaphorical. In *your* notebook, collect and draw the examples you find, assimilate them into your own repertoire of ideas... and, above all, experiment with and develop them – make them your own – in your design work.

INTRODUCTION
OUR WORLD OF SHADOW

This Notebook explores the architectural orchestration of shadow. Light illuminates the world but it is by shadow that we interpret what we see. When drawing on white paper – the artist's manifestation of ambient light – we conjure up our subject with grades of shadow. Without shadow, or with total shadow, there is nothing. We think of light and shadow as reciprocal opposites; but while light is general, shadow is localised and particular. When we think of architecture modifying light, it is always by making shadow that it does so. We architects enjoy drawing the appearance of shadow on paper or a computer screen but, more consequentially, our work when built makes and modifies shadow, real shadow.

A WORLD WITH NO SHADOW
Dhātu, James Turrell

In total light and darkness we are blind.

Dhātu, James Turrell, 2010

In 2010, in the Gagosian Gallery in London, the artist James Turrell created a room without shadow, filled with coloured light. It was entered through a square aperture at the top of a pyramid of steps. It was called 'Dhātu' (according to Buddhist cosmology a dhātu is a realm corresponding to a mental state of being). Passing through the square aperture you lost all sense of tangible surface (except for the slight reassurance that your feet were still touching a floor that you could not visually make out as such) and all sense of form. The space beyond the aperture was pervaded by a fairly bright even light. There were no edges, no corners, no points nor lines of reference by which you could tell where the space ended. Alone in there, there were no shadows. Though there was light you could see nothing. Only when a few other visitors entered could you see something – their ghostly shadowy forms.

James Turrell has created a number of similar installations in which our perception of architecturally finite space is dissolved or thwarted by the provision of ambient (and apparently source-less) light.

ANALYSING ARCHITECTURE NOTEBOOKS

' *"Are you there, men?" my eyes made out shadow forms starting up around me, very few, very indistinct.'*
Joseph Conrad – *The Shadow-Line* (1916).

It is by their shadows that we see things.

Shadows bring things into being.

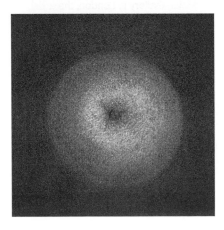

When other people entered Turrell's realm of even and undifferentiated light, you saw them by their shadows. To see something in this space of unrelieved nothingness was a relief. It provided something to relate to.

Shadow moulds our perception and understanding of the world we see. Though it is a prerequisite, it is not so much light that orchestrates our visual perception of the world but the appearance of shadow.

Shadow is the most important contribution to our visual understanding of form. With shadow things emerge from the polarities of light and darkness.

PHENOMENOLOGY OF SHADOW
in our interpretation of what we see

Shadows are insubstantial, ephemeral, unreliable.

Concluding a book on light in archaeology, the anthropologist Tim Ingold wrote of shadow:

'In the alternation of light and dark we enter, finally, into a world of shadows. Everything under the sun casts its shadow, upon itself, on other things, and on the earth. Yet unlike things, shadows have no substance, and they come and go. Out of doors they flicker with the breeze, as it brushes the surfaces of leaves and makes them tremble; indoors they flicker with the light-source, whether it be candle-flame or fire. They come and go, too, with the passage of clouds in the sky. Whenever the sun disappears behind a cloud, the shadow also vanishes. This is not – as is often thought – because the cloud blocks out the sun's rays, for were that so, every passing cloud would pitch us into black night! What happens in fact is that these rays are dispersed in all directions by atmospheric vapour, with the result that the illumination of surfaces is evened out. Thus, areas that were in shadow when the sun is out actually brighten up when the sun goes in. What fades is the contrast. Are things easier to see, then, in or out of the shadows? The default assumption, under a regime of modernity, is that shadows obscure rather than reveal. To see things as they really are, we insist, they must be brought out from the shadows. That is why we invest so heavily in the all-around, static illumination afforded by electric light, along with perfectly transparent glass and white walls. Medieval church-builders had different priorities, however. They were masters of shadow, of surface convolutions and dark corners, hiding things in alcoves and vaults in such a way that they would appear to emerge only with the shining of the light, through windows or from lamps, only to fade back into the woodwork or masonry once the light had passed. Whether shadows conceal or reveal depends, to an extent, on whether our interest is in the objective forms of things or in the textures of their self-shadowing surfaces. Texture shows up much better in a light that rakes the surface, coming at a shallow angle. Every crease or bump shows up in the contrast between the relative illumination of its light-facing facets and the relative darkness of facets in their shade. The weave of cloth, ripples of water, inscriptions in stone, blades of grass: all are picked out in these shadowy variations. Ever-changing with the light, shadows are ephemeral. To watch them is not to take the world in at a glance but to join in its temporal unfolding, almost as one would with an orchestral composition. It is to reveal a world that is not laid out in fixed and final forms, but launched in perpetual motion.'

We live in a world of light and shade…

'Deep shadows and darkness are essential, because they dim the sharpness of vision, make depth and distance ambiguous, and invite unconscious peripheral vision and tactile fantasy.'

Junichiro Tanizaki, trans. Harper and Seidensticker (1954) – *In Praise of Shadows* (1933), 2001.

'We find beauty not in the thing itself but in the pattern of shadows, the light and darkness that one thing against the other creates. Were it not for shadows there would be no beauty.'

Juhani Pallasmaa – *The Eyes of the Skin: Architecture and the Senses* (1996), 2005.

Tim Ingold – 'Commentary 1: On Light', in Papadopoulos and Moyes, eds. – *The Oxford Handbook of Light in Archaeology*, 2017.

OUT OF THE SHADOW
bringing life into the world

We make sense
of the world by
its shadows.

*We see things in light
but interpret them by
their shadows. Shadows
show us the textures and
forms of surfaces. They
might even lead us to see
strange creatures – such
as a hippopotamus –
emerging from the rocks.*

The world is visible to us in light. We think of light as the medium by which we see. But it might be said that we understand what we see more by shadow. It is possible to draw light – by applying white to a dark ground – but usually we draw by applying a dark medium to a light ground, i.e. by applying shade. This practice seems to be in accord with the way in which we interpret our surroundings by reading shadows. For example, you interpret the forms in the drawing above in a way that is reciprocal to that by which I depicted them – by the shade, nothing else; what you interpret as light is nothing but unmarked paper. By shadow you read the texture of the rock, its detailed topography, the depth of crevices, the gloom under the water's surface. It is by shadow too that you (and I) might see the things depicted in the shapes: even the head of an animal emerging from a crevice. Shadows are factors of perception.

By the flickering light of their tallow candles, people in prehistoric times saw hints of animals emerging out of the shadowy surfaces of cave walls. They tried to coax them from their rocky wombs with paint or by sculpting the rock to clarify their forms. So might animal herds, on which those people depended, be replenished. Shadows are harbingers of creation; they can be enhanced, intensified for creative purpose.

*This animal is on the
wall of a prehistoric rock
shelter in the Dordogne.
By carving, the shadowy
image of a horse,
suggested by features
already apparent in the
rock's surface, has been
enhanced. Acting as
midwife, the artist sought
to bring forth the animal
from its rocky womb by
intensifying its shadows.*

OUR SHADOW
companion, alter ego, playmate...

We play
with shadow.

We take our shadow with us wherever we go. For most of the time we do not notice it. Sometimes, in dramatic circumstances (left) or perhaps an art installation (e.g. 'Your uncertain shadow' by Olafur Eliasson, 2010*) or perhaps playing in the dim glow of a night light we do become aware of it and the ways in which it, instantly and exactly, replicates our movements. It is no wonder that shadow has been so extensively used as a metaphor for alter ego.

Shadow is implicated in the origins of art. Before the advent of digital sensors and printers, photography depended on shadow for the production of prints. The most primitive form of photography is the silhouette or stencil. Pliny the Elder, in his *Natural History*, tells of the origin of painting:

> 'The Egyptians assert that it was invented among themselves, six thousand years before it passed into Greece; a vain boast, it is very evident. As to the Greeks, some say that it was invented at Sicyon, others at Corinth; but they all agree that it originated in tracing lines round the human shadow.' **

And the story of Butades of Corinth:

> 'Butades, a potter of Sicyon, was the first who invented, at Corinth, the art of modelling portraits in the earth which he used in his trade. It was through his daughter that he made the discovery; who, being deeply in love with a young man about to depart on a long journey, traced the profile of his face, as thrown upon the wall by the light of the lamp. Upon seeing this, her father filled in the outline, by compressing clay upon the surface, and so made a face in relief, which he then hardened by fire along with other articles of pottery.' ***

'A shadow his father makes with joined hands And thumbs and fingers nibbles on the wall Like a rabbit's head.'

Seamus Heaney – 'Alphabets', 1987.

* See: olafureliasson.net/archive/artwork/ WEK100100/your-uncertain-shadow-colour (November 2019).
** Pliny the Elder – *Natural History* (c. 77–9 CE), Book XXXV, Chapter 5.
*** ibid. Chapter 36.

OUR SHADOWS
metaphor, and non-light shadows

Plato's 'Metaphor of the cave'

photography, television, cinema... depend on shadows cast

One of the most famous and influential uses of shadow as a metaphor is that proposed by Plato in his 'Metaphor of the cave', a discussion of how we perceive and interpret the world in which we live.*

Plato suggested that all many of us (the unenlightened) see of the world are the shadows of reality cast on the wall of a cavern in which we are chained.

All photography, television, cinema... (echoing Plato's analysis of how we only see the shadow of reality) depends on the projection of actual shadows onto a plain surface, or their electronic equivalent onto a photo-sensitive screen.

* See the *Metaphor* Notebook pages 118–19 and 129.

Light is not the only medium by which a shadow may be cast. Rain and pigment can leave 'shadows' that persist.

*A 'rain shadow' is the dry area left on the ground when, for example, we lie down in the rain. A tree, until the water seeps through its canopy, can cast a rain shadow. The artist Andy Goldsworthy has made rain shadows of his own body in various places.***

** See for example:
pbs.org/newshour/arts/watch-the-fleeting-beauty-of-artist-andy-goldsworthys-rain-shadows (September 2019).

And since ancient days we human beings have recorded our ephemeral presence in particular places by leaving lasting 'shadows' of our hands by spraying pigment over them as stencils.

SHADOW AS PLACE
a place to be

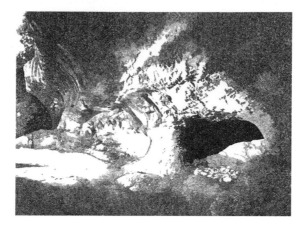

section

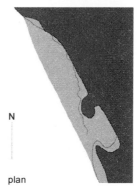

N

plan

The architecture of this prehistoric 'gallery' consisted in it being a place of morning and noon shade (above) that would become illuminated by afternoon and evening sunshine (top left).

Architecture is not only about seeing; it is about being. And a shadow can be a place to be.

The places where prehistoric artists chose to situate their work, coaxing life forms out of the solid matter of the earth, tended to be those that were themselves in shade, even total darkness. Deep caves contain depictions of animals that can only have been created by the light of flickering tallow candles. Other works were created under the shelter of rock overhangs. An example is the Abri de Cap Blanc (above) in the Dordogne region of France, which is where, some fifteen thousand years ago, the animal illustrated on page 11 was carved as part of a group of horses and bison. Here, together with their human creators, the carvings were sheltered from rain and from the hot summer noonday sun. This is the most architectural type of shadow: shadow as *place*. This is the type of shadow that protects us. This is the type of shadow in which we can hide. This type of shadow, when we occupy it, becomes a work of architecture.

Some of our most rudimentary works of architecture – such as when a ball-boy at Wimbledon plays the part of column supporting a roof in the form of an umbrella to protect a player from the heat of the sun during the break between games (right) – are concerned with the creation of shadow places. Shadow as place is an element of architecture. It is evident as such in permanent buildings, many of which act as 'umbrellas' or 'parasols' to shelter us from the sun and rain.

VECTORS OF ARCHITECTURE
directional conditions of architecture

Shadow is part of one of the world's fundamental vectors.

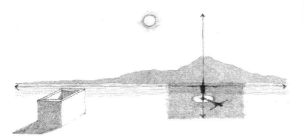

The sun and the shadows it casts constitute one of the primary vectors conditioning architecture.

In terrestrial space and time there are three primary vectors conditioning architecture. First is topography – directions implicit in the lie of the land that influence orientation and movement. Second is our own human form – the six directions (forward, backward, up, down, left, right*) implicit in our standing on the ground's surface. These are overlaid by a third vector: the directional light that comes from the moving sun. A challenge for architecture is to find an accommodation for all three, even though they are not always aligned. Shadow is the reciprocal of the light vector, with its own architectural possibilities and problems.

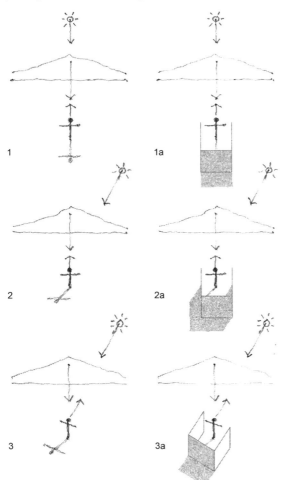

The fundamental relationships between the three primary vectors of architecture are illustrated in these sketches (left). The distant mountain represents the directional pull of topography. The figure with outstretched arms represents our human form with its innate directions. And the sun represents itself and creates its reciprocal shadow. In 1 these three vectors are in resonant alignment. In 2 the sun is out of alignment while the figure and mountain remain aligned. In 3 the figure has rotated out of alignment with the mountain and into alignment with the sun.

Architecture (right: 1a, 2a, 3a) mediates between these three vectors.

* See *Analysing Architecture*, 4th ed., 2014, pp. 144–8.

CIRCADIAN ARCHITECTURE
responding to diurnal rhythms

Shadow is part of architecture's diurnal rhythm.

The word 'circadian' is often applied to our personal biological rhythms: how some of us might work best in the early morning and others late in the evening; how people generally might be predisposed to a doze after lunch; and so on. Jet lag disrupts your circadian rhythm. Architecture has circadian rhythms too. Conditioned by the passage of the sun across the sky, the cycles of day–night and the seasons, as well as by how our daily routine relates to our use of space, architecture is not settled to some generic present but relates to rhythms of time. Shadow plays its part in the circadian rhythm of architecture. The sequential drawings on the right simplify the diurnal rotation of shadow in the northern hemisphere.

morning (external)

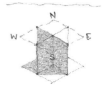

morning (internal)

noon

noon

evening

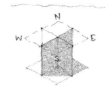

evening

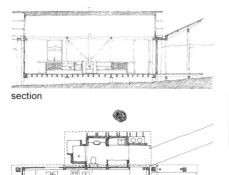

section

plan

→ N
direction of noon sun in Australia

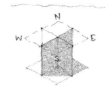

Glenn Murcutt's Kempsey Guest Studio* (1992) might be classed a circadian dwelling. Under the shade of a tree (above), each of its sides responds in its own way to different times of the day.

* See *Twenty-Five Buildings Every Architect Should Understand*, 2015, pp. 147–52. Also: atlasofplaces.com/architecture/guest-studio/ (October 2019).

SHADOW TYPES

As architects, we should be aware of the different ways shadow can be an ingredient in our particular and general art. Leonardo da Vinci categorised how an artist sees shadows. His classification includes three types: the shadow cast by an object; the shadow attached to an object (its dark side); and shading (the gradations of shadow on a curved surface). His concern was with the depiction of objects (faces, bodies, things…) in two dimensions, i.e. in drawing and painting. In architecture, which deals in space and matter – in time, real light and the actual world – there are more than these three types of shadow.

BASIC SHADOWS
'attached', 'cast' and 'shading'

attached shadow/dark side

shading   cast shadow

We all know what shadows are. Any object standing in directional light creates an area of darkness (to the side facing directly away from the source of light) where the light is obstructed. The shape of this shadow is determined by the shape of the object.

There are three basic kinds of shadow. First, the body itself prevents the light touching its side facing away from the light – its dark side. This is 'attached' shadow. (Leonardo da Vinci, opposite, called it 'primary' shadow.) Second, light is obstructed in the space behind the object. This is 'cast' shadow. (Leonardo called it 'derived' shadow.) It is cast shadow that projects the dark patch on the ground that follows us around on a sunny day, which we would point to if asked to indicate our own shadow. There is a third kind of shadow – 'shading' – where we see a gradation of shadow on surfaces at an angle to the source of light.

These three kinds of shadow can be seen in the drawing above. They are all caused by the obstruction or reduction of the light striking surfaces. They play their part in our understanding of the world. By comparison with the line drawing alongside (above left) we can see that it is by reason of the attached shadows and shading that we gain a stronger perception of the form and mass of the object. And it is by reason of the cast shadow that the object appears grounded; i.e. the shadow makes the position of the manikin on a table top apparent. But generally, it is because of the shadows that we see the object in the drawing as standing in light.

These three kinds of shadow are defined primarily in relation to visual perception and depiction – drawing, painting, photography… Though these are not irrelevant in architecture, there are other ways of categorising shadow.

LEONARDO'S SHADOW TYPES
in drawing and painting

Leonardo da Vinci made extensive notes on shadow in his notebooks. He wrote:

> 'No substance can be comprehended without light and shade; light and shade are caused by light.' *

He identified many subtle types of shadow but all, he noted, originate in two: 'primary' and 'derived':

> 'Primary shadow is that side of a body on which the light does not fall. Derived shadow is simply the striking of shaded rays.' *

These are the shadows on the surface of a body and shadows cast by the body on other surfaces.

Leonardo also acknowledged gradations in shade on curved surfaces:

> 'The beginnings and the ends of shadow extend between light and darkness, and they may be infinitely diminished and increased.' *

As a draughtsman, Leonardo da Vinci's primary concern was with perception, and pictorial depiction on a two-dimensional surface. In his drawing (above) he renders the appearance of form with gradations of tone. It is composed mainly of 'primary (attached) shadows' and 'shading'; but there are also small 'derived shadows' cast by the nose and eyelashes.

* Leonardo da Vinci (15th century CE), trans. MacCurdy, 1939.

ARCHITECTURE'S SHADOW TYPES
line drawing and shading

Shadows are essential to our interpretation of the world.

Architecture is not only about depiction. .'. is about intervening in, changing, the real world. Shadows in architecture are actual – cast by real light in all its variations – rather than pictorial. (I can only show them pictorially here but please imagine them in reality.) Leonardo's three type shadow classification – 'primary', 'derived' and 'shading' – is apparent in architecture but it is not sufficient.

These two pages illustrate those three types of shadow as they apply to the example of a simple gable wall built of large stones.*

Architecture is often depicted in line drawing. Even though perspective gives an illusion of three dimensions, there is no light; there is no shadow.

'SHADING'. The texture of the wall is shown by the stones' shaded contours even on an overcast day when the quality of light is diffused.

* The gable wall belongs to a small cottage called Llainfadyn re-erected at the St Fagans National Museum of History near Cardiff. This slate-worker's cottage is also the subject of Case Study 3 at the end of *Analysing Architecture*, 4th ed., 2014.

ARCHITECTURE'S SHADOW TYPES
'primary' and 'derived'

What we see is different types of shadow layered over each other.

'PRIMARY' SHADOW. The surface of the wall is in shadow when it is facing away from the sun. It also casts its own 'derived' shadow.

'DERIVED' SHADOW. The shadows of adjacent trees, with patches of sunlight striking through, add an extra layer of light and shade to this gable wall.

Leonardo's types of shadow play their parts not only in the ways we see architecture but also in how we use and inhabit it. A 'derived' shadow can be a place to be or avoid: a derived shadow can be a good place to be on a hot day; but it would not be a good place to try to grow sun-loving plants. A 'primary' shadow across the elevation of a building gives it a different character to when it is illuminated by bright sunshine. And it is by 'shading' that we read the form of a building.

But, as I have said, there are more than these three types of shadow in architecture. They can be categorised in a different way by focusing on architecture's primary defining motivation – identification of place. These categories are illustrated on the following pages.

ARCHITECTURAL TYPES OF SHADOW
shadows for occupation and passing through

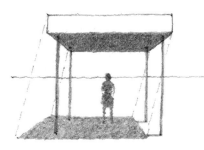

SHADOW CONTAINER

This is a 'derived' shadow that contains. In this image the person depicted is contained by the shadow cast by the canopy. That becomes the room of shade within which the person stands even though it has no tangible walls. A 'wall' of this 'room' only becomes apparent when, for example, you put your hand through it to become illuminated by the sun. With a parasol you carry around with you your own shadow container.

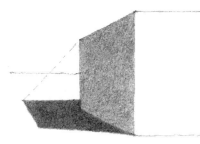

DARK SIDE

This is an attached or 'primary' shadow. But it is more: its association with a cast or 'derived' shadow makes it a place. In the northern hemisphere it is the north side of a building; in the southern, the south. In the morning it is the west side; in the evening the east. The dark side has a particular character, sometimes thought to be negative. But it can be used in positive ways too.

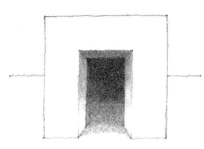

CONTAINED SHADOW

This is the shadow that is contained within an enclosed room, a room with walls as well as a roof. The inner wall surfaces are its impenetrable edge. (See the Forster quotation on page viii.) If a door was fitted and closed, the contained shadow would be complete darkness. Nothing (except maybe phantoms) would be visible in such intense shadow. The darkness we see through the doorway of the cottage on page 27 is contained shadow.

SHADOW THRESHOLD

This is a shadow line that is crossed when passing from one place into another. In the instance illustrated (left), the shadow threshold is created by the doorway and its lintel. The contained shadow in the doorway (above and on page 27) also has its threshold. When entering or exiting a doorway we are usually also crossing a shadow threshold.

ARCHITECTURAL TYPES OF SHADOW
shadow types that contribute to visual interest

Shadow of course plays a role in the visual appearance of architecture too. Through the day shadows move as the sun crosses the sky. Their changing shapes and patterns counterpoint the fixed material form of built fabric. Shadows are essential ingredients in the orchestration of contrast between light and dark. They contribute to our perception of three-dimensional form; it is by shading that we see a column as being circular in plan; it is by cast shadow that we read the depths of recesses. We can draw images on walls with incisions and relief mouldings that produce shadows. Walls and floors can be screens for the projection of shadows. Shadows can frame views, and evoke infinite depths...

Shadows in the pictorial world of drawings and paintings stay fixed. In architecture shadows move and change. They move as the sun crosses the sky and with the seasons. Shadows change with the weather: breezes rustle leaves, clouds scud, overcast skies soften light... Our own shadows move as we do.

SHADOW FRAME (right)

When there is a hole in a wall (like the window illustrated here), and outside is light whilst inside is dark, then, as in this drawing, the view through the hole (window) is framed by shadow. The shadow intensifies the 'picture' through the window. The brightness of the light from the window intensifies the shadow that surrounds it. (See page 29.)

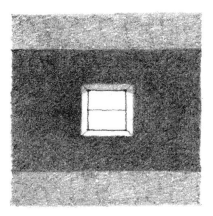

The reciprocal of the shadow frame is the framed shadow (below).

FRAMED (DEEP) SHADOW

Shadow seen through a hole in a brightly lit wall – into contained shadow – can appear infinitely deep. The darkness of the shadow is intensified by the lightness of the wall. Inside is mystery.

FRAMED SHADOW FRAME

A framed shadow may also be a shadow frame. A framed shadow is in the background but it can also be a shadow frame for something in the foreground, e.g. this bust.

SHADOW TYPES
visual interest

PROJECTED SHADOW

A surface – the ground or (in the illustrated instance) a wall – can be a screen on which ' derived' shadows are projected. The surface of a plain wall can be enlivened by the moving shadows of the leaves and branches of an adjacent tree. Sunlight streaming through a window can cast a shadow of the window's frame onto the floor. Shadows projected by the sun move during the day as the sun moves across the sky.

SHADOW GRADIENT

Leonardo pointed out the variations in shade on curved surfaces (page 19). But shadow gradients in architecture can be created in different ways. In the drawing (left) light from a source hidden by the upper (dark) section of wall is washing down a wall creating a shadow gradient from light to darker.

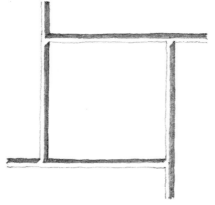

DRAWING WITH SHADOW

Drawings can be made on the surfaces of walls by incisions and relief carving. The image or pattern is made visible by the shadows cast. This technique can be used to decorate walls with patterns and ornament. But it can also be used to accentuate joints between materials (as in recessed joints between bricks or stones) or to frame openings with mouldings (see page 47).

EXAMPLE: SHADOW CONTAINER
shadow as a place to be

A shadow container
is a place to settle.

We might want to bask in the sun whilst also protecting our children from it. The architectural purpose of a parasol is to provide a shadow container, a shadow we can occupy (or not).

A parasol is a simple work of movable architecture. It provides a temporary shadow container.

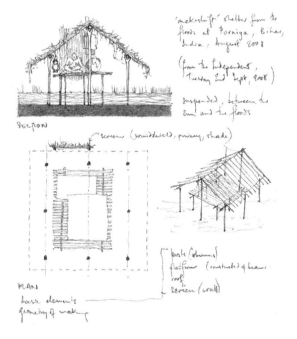

I sketched this shelter from a report in a British newspaper on the 2008 floods in India. This simple yet subtle work of architecture provides respite from the floods with a platform lifted out of the water. But its roof also creates a shadow container providing refuge from the strong Indian sun.

EXAMPLE: DARK SIDE
the shadow side

A house in sunlight is welcoming. A 'dark side' can be forbidding.

Light and dark sides have different characters.

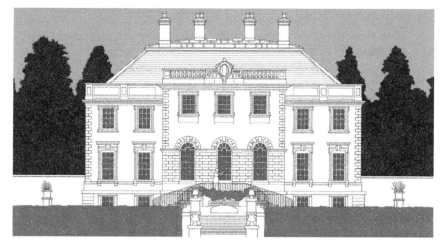

Objects under directional illumination have a light side and a dark side. It is the same with buildings. During the day we are lit by the sun while during the night we are on the earth's dark side. In the northern hemisphere, during the day, the south-facing side of a building (above) receives more light than the north-facing side (below). In the morning the west-facing sides are the dark sides; in the evening the east. The dark, the side shadowed, has a different character from the light or sunlit side. The orientation, hence the relative shadowing, of different parts of a building affect how we experience and relate to them. A dark side presents a different, perhaps more daunting, character compared to a light side.

The drawings show the sunlit south elevation and the shaded north elevation of William Adam's House of Dun (finished in 1743). (See also pages 102–4 for illustration of how the house's orientation, and hence shading, affects how it is experienced and what that experience tacitly says about the owner's situation.)

ANALYSING ARCHITECTURE NOTEBOOKS

EXAMPLE: CONTAINED SHADOW
a view into a dark interior

A 'contained shadow' can be inviting, or daunting.

In total light and darkness we are blind.

Works of architecture are more than pictorial or sculptural forms. We experience them as places, places in which we live, work and enjoy ourselves.* Shadow is an ingredient in place-making. Along with temperature, sound, scale... it is one of the modifying elements of architecture;** and it is the reciprocal alter ego of another of those modifying elements, light.

One of the most powerful of the ways shadow contributes to place-making – the contained shadow – is illustrated by the dark doorway in the image above (which also shows different cast shadows). Architectural shadows are different from drawn shadows in that we, in the actual world rather than in pictorial depiction, may pass into them and occupy them. They are shadows that are habitations. In the case above, the contained shadow seen through the doorway is either off-putting – if we are a stranger – or welcoming – if we are coming home.

Attached and cast shadows, as well as shading, are apparent in the above drawing of a cottage. But the most powerful shadow is the deep contained shadow framed by the doorway which marks its threshold. This identifies a place we can enter, occupy, inhabit... It is a shadow that invites or repels.

* See *Analysing Architecture*, 4th ed., 2014, pp. 25–34.
** Ibid., pp. 47–60.

EXAMPLE: SHADOW THRESHOLD
shadows we pass through

> Shadow thresholds intensify the experience of entrance.

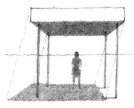

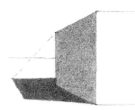

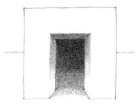

All shadows have their thresholds (left) but the experience of entering – for example, a bright sunlit garden – can be intensified by first passing through the shadow threshold created by an archway.

The shadow threshold is implied in figures of speech, e.g. 'the darkest hour is just before dawn'. In Psalm 23 (see page 1) it is implied that the 'valley of the shadow of death' is the fearful darkness that precedes entry into the celestial city of light. In the architectural orchestration of light and shade, whether metaphorical or real, the shadow threshold is an essential device.

All shadows have their thresholds. Sometimes these are sharp and we pass from light into shade in an instant (above). Sometimes the edges of shadow are vague and passage from light into darkness is gradual. The third kind of shadow threshold (above right) is perhaps the most dramatically effective. It provides a moment of shadow as a prelude to emerging into brightness. Thereby the brightness is intensified by its contrast with the preceding shade.

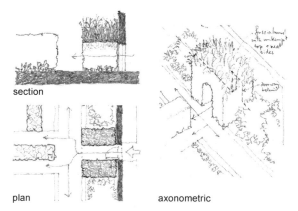

section

plan

axonometric

The walled garden at the Castle of Mey (Scotland) has a shadow threshold. You pass through a doorway and a deep dark fuchsia tunnel before entering the bright and colourful garden.

As a child we were taken to beaches on the Gower Peninsula in South Wales. The route was along the coast of Swansea Bay and then up a hill and over a common. But the moment when I felt we had entered the bright 'other world' of the Gower beaches was when the road passed through a dark tunnel of overhanging trees. This is perhaps my earliest memory of the emotional power of a shadow threshold. It is still there.

EXAMPLE: SHADOW FRAME
a view from shadow into light

A shadow frame can dominate the character of a room.

Shadow frames are related to thresholds.

The character of a room can be dominated by a shadow frame. The quintessential example is in a traditional theatre auditorium where the brightly lit action on the stage is framed by the darkness of a proscenium arch. Or in a church where a brightly lit sanctuary, containing the altar, is framed by a dark archway. Shadow frames play their part in domestic architecture too. Most daytime views out of windows are framed by shadow. And dark spaces can lead into light ones with the threshold framed in shadow (as in the above image).

Here the warm brightness of a conservatory is seen from the cool shade of a living room. The shadow frame is a major contribution to the character of the room. (Note too that the cat is crossing a shadow threshold.)

EXAMPLES: NARRATIVE SHADOW
shadows that tell stories

Doorways present different shadow types and relationships according to different lighting conditions. A door may be open to a bright corridor but with the room dark; the room dimly lit with the door open to a dark corridor; light squeezing from the corridor through the cracks around the closed door; and a line of darkness around the door with the room lit. Other permutations are possible. Each tells a story about relationships. They imply narrative.

As a 'side-light'… shadow types contribute to the narratives of architecture.

'All thought flies' reads the inscription above the mouth – a framed deep shadow – of The Ogre's Head (The Mouth of the Underworld) in the Italian gardens of Bomarzo. Imagine the change in narrative if that mouth were filled with light (or roaring flames) rather than the empty darkness of deep shadow.

SHADOW FRAME: door wide open; room light off; light in corridor outside on. The contained shadow in the room becomes a hiding place from which the corridor can be watched… like a stage. (See Kafka image opposite.)

FRAMED SHADOW: door wide open; (rather dim) room light on; light in corridor off. Someone passing by along the dark corridor would be tempted to look into the lit room… as into a display case.

EXAMPLES: NARRATIVE SHADOW
shadows that tell stories

Novelists evoke shadow images to enrich their narratives.

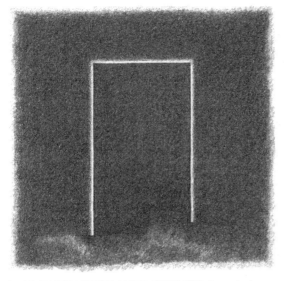

SHADOW FRAME/FRAMED SHADOW: door shut; room light off; light in corridor on. The room is contained in its own darkness, but with a sign – the light on – that the corridor is still 'alive'. (See Kafka image right.)

FRAMED SHADOW FRAME: door shut; light in room on; light in corridor outside off. (I created this image by inverting the one above.) The room is contained within its self, its own dim light.

The relationship between shadow and narrative can be reversed. Authors evoke images of shadow types in their words as support for their stories.

K. waits in the dark of his room waiting for a fellow boarder to return home.

Light under a fellow boarder's door indicates that he is at home.

Franz Kafka evoked images of shadow frames and framed shadows in the narrative of his novel The Trial (Der Process, 1914–15; 1925), 1937. I have sketched two of many instances here.

EXAMPLES: CAST SHADOW
shadow play on a wall as screen

In his 'Metaphor of the cave' (see page 13) Plato suggested that all the unenlightened see of the world are shadows projected onto the wall of a cave. All photography – still or moving, chemical or digital – depends on the projection of shadows onto a screen. Projected shadows play a part in our experience and perception of the world. Projected shadows contribute to the aesthetic joy of architecture too. The can be planned but also happen by chance.

The shadows cast on the walls, floor or ceiling of a room by an open doorway, the glazing bars of a window in sunshine or the reflection of sunlight off the rippling surface of water, lend a room vitality it would not otherwise have. A room totally in shade can be a dull room. The admission of sunshine, recognised by the shadows it casts, brings a room to life. The sunlight intensifies the shadows and creates gradations.

Sun striking through the bare branches of winter trees give a white wall a dynamic 'wallpaper'. The shadows move with the breeze and the sun. They fade and reappear as clouds intervene. The nature of the shadows changes with the seasons of the year as buds turn to leaves in the spring, flourish in summer, and then, in autumn, fall to the ground. They tell an annual story. It is a movie for which architecture can intentionally provide a screen.

Standing alone, a tree's form is profiled against the sky, its shadow unnoticed on the ground.

Place the tree next to a sun-facing wall and it acquires a companion – its shadow.

EXAMPLES: SHADOW GRADIENT
shadow and depth

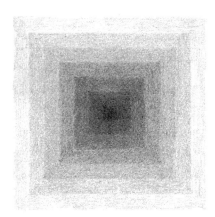

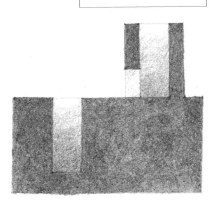

In drawings shadow gradients give a sense of depth. Shadow gradients are often associated with deepness. They occur at the mouths of caves and of tunnels where one is aware of a crescendo of shadow reaching into a dense darkness that eventually melds with the solidity of the rock.

Shadow gradients are associated with verticality: light towards the top and dark towards the bottom. They are the shadows of a well. They are the shadows of the faint light from the sky you might see drifting down a wide open chimney onto a cold grate below.

Early Minoans on the island of Crete cut tombs deep into areas of exposed rock. They were entered down steps into a slot in the rock leading to the tomb doorway.

Descending you go down through a shadow gradient to the framed shadow of the doorway. Returning you ascend through the same gradient to the light.

EXAMPLES: DRAWING WITH SHADOW
from prehistoric pattern-making to defining place

Blombos Cave rock art, 70,000 BCE

For tens of thousands of years we have made marks by scratching lines onto surfaces. We see those lines by shadow. As soon as we scratch a line in the sand or on the surface of a rock we enlist the power of shadow. Effectively, since (all other things being equal) neither the sand nor the rock change colour, when we inscribe a line we are drawing with shadow. Such drawings might be decorative or symbolic. But since footprints or a line in the sand (left, middle and bottom), as well as lines of footsteps (above), all identify place, then, as well as depicting patterns and images, drawn shadows can be primary elements in the creation of architecture.*

* See for example, pages 136 and 137 of the *Children as Place-Makers* Notebook.

Drawn shadows – the shadows of scratches, engravings, carving… – can provide surfaces with decoration (above left) and with images (left, top), which may or may not be symbolic. But by identifying a place with footprints (left, middle), a pathway with steps in soft sand (above right) or outlining a circle of place (left, bottom) – all of which manifest and accommodate human presence – drawn shadows can also be architecture, instruments by means of which we scribe lines of demarcation and lines of movement onto the ground.

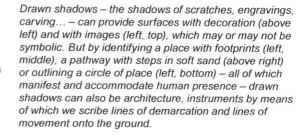

DRAWING SHADOW
SCIAGRAPHY

With thought and consideration, as architects we can orchestrate all the types of shadow described so far in this Notebook to enhance and enrich the experience of real buildings. But shadows have also played a part in architects' predilection for making their work look good in drawing, as a way of impressing clients. We use drawings to develop, explore and illustrate architecture. A drawing might consist of simple lines; but to give it depth, and a heightened impression of three-dimensional reality, shadows are an effective elaboration.

The practice of geometrically constructing drawn shadows occupied a substantial proportion architectural students' studio time during the late nineteenth and early twentieth centuries. It has its own name: sciagraphy, derived from *skia*, the Greek for shadow. Whole books were devoted to providing guidance on this essential graphic technique. Some examples are given on the following pages.

POWER OF SHADOW IN DRAWING
without shadow:

The drawings on these two pages show the power of shadows (sciagraphy) to dramatically increase the three-dimensional appearance of architectural drawings.

A line drawing appears flat…

CONJURING THREE DIMENSIONS
with shadow:

… and makes
a drawing more
arresting.

*… but shadows give it
depth and mass.*

Shadows reinforce the illusion of reality in a drawing.
They also make the drawing more aesthetically
engaging.

TONAL VARIATION
skill in laying watercolour washes

McGoodwin, 1904; drawing by G.F. Stevens.

Beaux-Arts architects (towards the end of the nineteenth and into the twentieth century) were concerned with the appearance, ornamentation, proportion and modelling of the surfaces of buildings. They prized the production of drawings made engaging by their illusion of three dimensions. Part of the skill was to use carefully laid watercolour washes to give those drawings depth with shadow. The drawing above illustrates the subtle use of graded tones and the different ways those tones were used to convey depth, modelling and detail.

The opening in this drawing illustrates 'framed shadow'. Generally, Beaux-Arts architects were mostly concerned with 'cast shadows' and 'shading'. Some 'drawing with shadow' is evident too: e.g. the 'key frieze' along the balcony edge.

SELLING ARCHITECTURE
shadows in Beaux-Arts drawings

Architects need to market their designs.

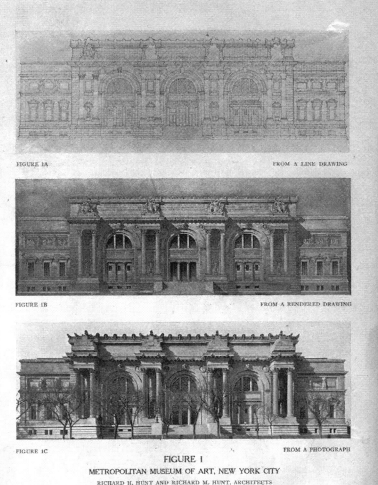

FIGURE 1A   FROM A LINE DRAWING

FIGURE 1B   FROM A RENDERED DRAWING

FIGURE 1C   FROM A PHOTOGRAPH

FIGURE 1
METROPOLITAN MUSEUM OF ART, NEW YORK CITY
RICHARD H. HUNT AND RICHARD M. HUNT, ARCHITECTS

McGoodwin, 1904.

'It is with shadows that the designer models his building, gives it texture, "colour", relief, proportions. Imagine a building executed in pure white marble and exposed, not to sunlight, but to uniformly diffused light that would cast no shadows. The building would have no other apparent form than that of its contours. It would seem as flat as a great unbroken wall. Cornices, colonnades, all details, all projections within the contour lines, would disappear.'

Henry McGoodwin illustrated the point he made in the quotation (left) with the above images. Sciagraphy was (is) part of architectural salesmanship.

Henry McGoodwin –
Architectural Shades and Shadows, 1904.

GEOMETRY OF CAST SHADOW
calculated projection

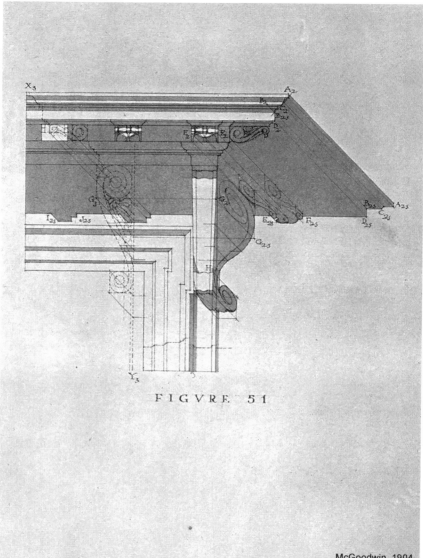

FIGVRE 51

McGoodwin, 1904.

Sciagraphy, the rendering of shadows, could be determined by geometry. If parallel rays, from an imaginary sun, struck the drawn element at 45° (in both dimensions) then the length of shadows cast would correspond directly with the depths of projections. The result would be a drawing accurately depicting depth. You can see the relevant construction lines above.

In the above drawing the dimensions of the shadows are geometrically determined by those of the projections.

ANALYSING ARCHITECTURE NOTEBOOKS

GRAPHIC EFFECT
illustrating depth, but not reality

Sciagraphy is not real.

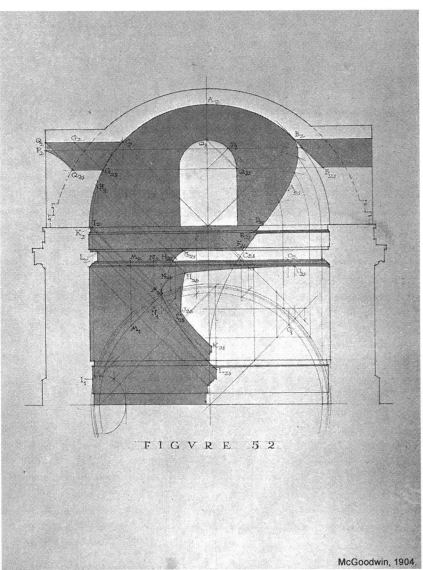

FIGVRE 52

McGoodwin, 1904.

The accurate depiction of shadows, especially of curved forms, could involve complex geometric construction. Sometimes, as in the drawing above, the effort involved was more for graphic effect and dimensional communication than the depiction of how a building or space would appear in reality. The domed space drawn in section above would never be lit as shown.

Depth in recesses could be drawn too, sometimes involving intricate construction lines.

A GLASS BEAD GAME
a graphic conceit

Drawing shadows
can become a fetish.

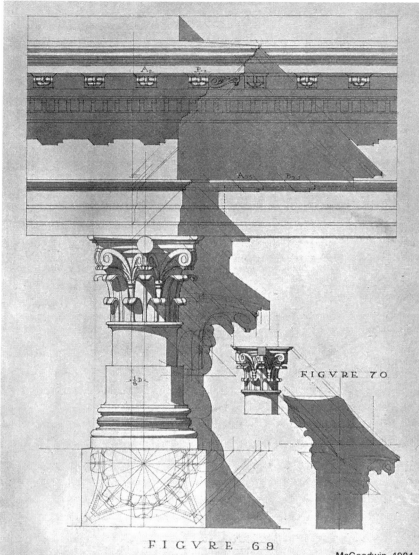

FIGVRE 70

FIGVRE 69

McGoodwin, 1904.

Though such Beaux-Arts sciagraphy was intended to
increase the accuracy of the depiction of proposed
building designs, its emphasis on drawn 3D effect
did not give a full account of the potential of shadows
in architecture. This purely graphic use of shadow
reinforces the limiting idea that architecture is primarily a
visual art (rather than a rich and complex frame for life).

See also, for example:
John M. Holmes – *Sciagraphy*,
1952.
P. Planat – *Manuel de
perspective et tracé des
ombres*, 1899.

THE PLEASURE OF SHADOW
taking pride in drawing

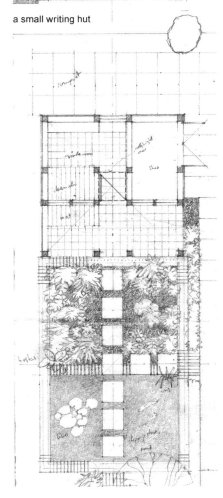

a small writing hut

The fine quality of the drawings produced by students during the Beaux-Arts years of architectural education suggest that, though it was probably hard and meticulous work, they gained great satisfaction in producing them. The acquisition and application of skill in drawing and the laying of watercolour washes, along with the convincing appearance of three-dimensional depth achieved with sciagraphy, must certainly have contributed to a sense of creative pride.

In my own design (on this page and the next), I enjoy the sensual enjoyment of the act of drawing as well as the intellectual stimulus of the challenges involved. This includes the pleasure of applying graphite to an appropriate quality of paper. The illusion of three-dimensional form achieved by the application of shadows provides its own peculiar satisfaction too. Shadow contributes to what one might call the magic of architectural drawing.

a local government HQ

The application of shadows to create the illusion of three-dimensional depth is part of the sensual pleasure of drawing.

GRAPHIC SHADOW
giving drawings character

Subtly different shadows give drawings different atmospheres.

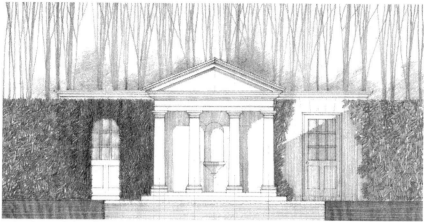

swimming pool changing rooms

In your own drawings you can try out depicting shadows in different ways. Sciagraphy is not just a geometric science, it is a technique open to experiment. The two drawings above are obviously of exactly the same proposal – changing rooms for a swimming pool in the grounds of a substantial neoclassical country house near Bath – but the atmosphere of the drawings is subtly different due to the use of different media and methods of showing shadows.

In the drawing of a similar subject on the chapter title page opposite I have depicted shadow in another way. As you go through this Notebook you will find various ways in which shadows can be rendered in architectural drawings.

Subtle differences in the depiction of shadows in architectural drawing can affect the character of the atmosphere evoked.

DRAWING WITH SHADOW

Light is the white page; pencil marks make the shadows by which the image is revealed. But in actual (rather than drawn) architecture the relationship is real: light is light and shadow is shadow. And one way in which we can use shadow is to mould or carve the surfaces of our buildings to 'draw' lines, patterns and images in relief. This, as distinct from drawing shadow (as illustrated in the previous chapter), is drawing *with* shadow. As illustrated on page 11 this is something we human beings have done since prehistoric times. We see a surface and want to scratch our names, symbols and images into it, using shadow to make drawing. The same principle has informed ornamental pattern and image making on the surfaces of architecture through the centuries.

45

DRAWING WITH SHADOW
engraving

We have been engraving images and patterns into surfaces through all history. Sometimes we scratch graffiti into a rock wall or tree trunk to record our presence in a particular place. We can make more formal art and decoration too.

prehistoric rock engraving, Libya

In prehistory we scratched the images of animals onto rock walls. The process became more refined in the hieroglyphic imagery of ancient Egypt (right) and the decorative low relief friezes of Greek and Roman architecture (below).

marble base, Athens

hieroglyph carved into granite obelisk

All historical periods of architecture have decorated their surfaces with images and patterns drawn with shadow. Below is an example from the Secession in Vienna.

Ancient Egyptians created images on flat surfaces with the shadows around the edges of carved depressions. This is a hieroglyph of the god Horus.

Secession Building, Vienna, Joseph Maria Olbrich, 1898

USING SHADOW LINES AND GAPS
to frame and separate

Shadow delineates.

Shadow is used in architectural detailing. For example, traditional windows often have mouldings to their glazing bars (below) that catch shadows (right), framing the view through the glass with parallel lines that soften the edge of its shadow frame.

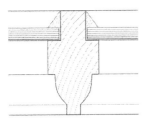

ovolo moulding

traditional small paned window with ovolo mouldings

The frame shown has ovolo mouldings (above), the most common moulding in traditional small-paned windows.

Above: the linear shadows of the mouldings help soften the edge of the light coming through the window.

Below: shadow gaps reinforce the separation of construction elements; here between a column and roof.

We use 'shadow gaps' to give an appearance of constructional clarity. A shadow gap might be placed between two materials to draw a sharp line between them. In his Secular Retreat (2018), Peter Zumthor designed a shadow gap between the rammed concrete columns and the slab-like roof (right). The gap helps to distinguish the elemental identity of each and also lightens the apparent weight of the roof by suggesting that it is not quite supported by the columns. (The actual support is hidden in the shadow.)

Secular Retreat, South Devon, Peter Zumthor, 2018

SHADOW AND PLASTIC FORM
formal and aesthetic effects

Doric columns, the Propylaea of the Athenian Acropolis

The fluting of Doric columns creates shadows that emphasise the columns' roundness and their transmission of vertical forces (above).

Even out of direct sun the soft shadows of a column's mouldings inform our reading of its form and enhance its aesthetic appearance (below).

Ionic column base, the Erectheion on the Athenian Acropolis

The columns of ancient Greek temples were often carved with vertical fluting. The shadows the arrises cast on the columns' surfaces, progressing from narrow to broad, enhance their appearance of solidity and roundness as well as accentuating their role as vertical structural elements.

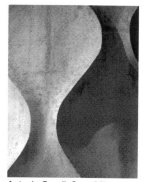

Antonio Gaudi, Casa Mila, 1912

Through history architects have moulded details of their buildings to catch shadows. Such details range from the regular and geometric to more sculptural irregular forms. Such sculptural mouldings create soft shadows even out of direct sunlight.

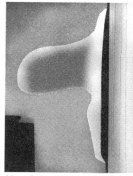

Alvar Aalto, Villa Mairea, 1939

GRAND MOULDING
Bagsværd Church, Jørn Utzon

Moulding form to catch and modulate shadows can inform whole buildings too. Jørn Utzon's Bagsværd Church in Denmark was completed in 1976. It is a building conceived from the inside out. The church's interior has a curvaceous ceiling (right) designed to modulate light and shadow in a way inspired by clouds Utzon had seen in Hawaii.*

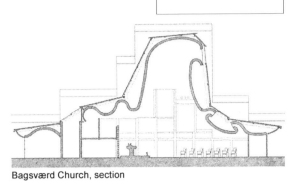

Bagsværd Church, section

Utzon's ceiling in the Bagsværd Church (below) is an architectural instrument for grading shadow to evoke the idea of rolling clouds in the sky. This instance of using shadow to mould form dominates the space.

* See also page 134 of the *Curve* Notebook, and Richard Weston – *Utzon*, 2002.

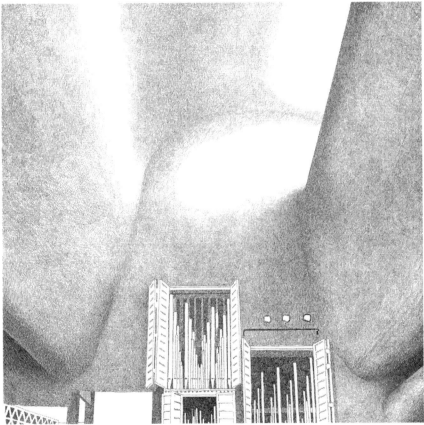

Bagsværd Church, interior

SCULPTURAL MODELLING
shadow gradients on curved surfaces

Guggenheim Museum, Bilbao, Frank Gehry, 1997

Moulding and modelling contribute to the sculptural possibilities of architecture by the ways they grade shadow. They enhance the aesthetic interest of both external and internal form with shadow gradients: light swells and shady troughs.

We see curvaceous architecture as sculptural. The subtle variation in shading on its curved surfaces contrasts with the hard-edged shadows of orthogonal buildings. They reflect and disperse light as part of their aesthetic power.

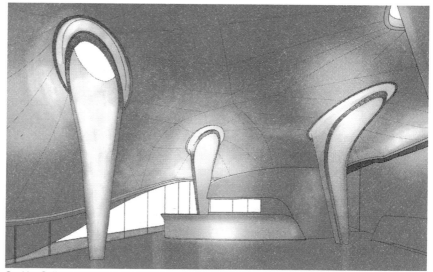

Sackler Gallery, London, Zaha Hadid, 2013

PRACTICAL SHADOW

Shadow can have a practical purpose but that purpose might be different in different circumstances.

I once organised a lecture for second year students on the use of a piece of software for predicting the environmental performance of interior spaces. Shading was one of the factors the software modelled. The lecture was presented by a newly appointed colleague (who had written the software) recently arrived from Australia. (We were in Wales.) He set about illustrating the modelling of sun paths and their effect. His main intent, it became apparent as he spoke, was to prevent sunlight entering through the windows of the building... completely. I had to point out to him that, having lived in Queensland, I knew that keeping the sun out of Australian rooms was a good idea because it would ease the need for air conditioning by reducing solar heat gain. But, I continued, we were in Wales! And that, on those rare occasions when the sun was shining, we were quite happy to welcome it into our rooms.

The practical benefits of shadow are different in different climates.

PRACTICAL SHADOW
how shadow might be used

1 tree shading a house

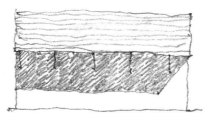

2 overhanging eaves

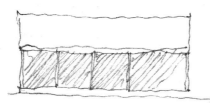

3 veranda

There are various practical ways in which we use shadow to keep cool in warm climates. Here are some examples:

1 Use a tree to shade a house and reduce solar warming. (See the Murcutt house on page 16.)

2 Give a roof a deep overhang to shade walls, helping keep them and the building's interior cool.

3 That large roof overhang would create a relatively comfortable place, so make it into a veranda.

4 Lift a house on stilts or stumps, with a shady and comfortable place underneath.

5 Build a shady loggia on the roof of a building where it will also catch cooling breezes.

6 Make streets narrow so the sun cannot reach into them.

7 Deepen the shade of those narrow streets with overhanging eaves.

8 Construct a parasol, an architectural shade structure.

9 Shade a courtyard (patio) with fabric drawn across cables.

10 Plant a tree to shade a courtyard.

11 Provide windows with awnings to keep out the sun.

12 Hang shade cloth across a balcony when needed.

These are shading strategies that have been used for centuries by people with no architectural pretensions because they help make life more comfortable. All architects can learn from their modest predecessors.

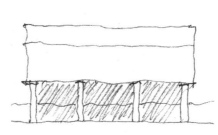

4 undercroft

5 roof loggia

PRACTICAL SHADOW
how shadow might be used

6 narrow street

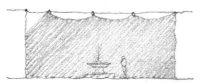

7 narrow street with overhanging eaves

8 built parasol

9 shade cloth across courtyard

10 tree shading courtyard

11 shade or awning over window

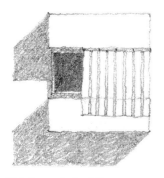

12 balcony shade cloth

There are many more instances than those shown in these quick sketches. Even in cooler climates some of these ways of creating and using shade can make for pleasant places to be. It is worth keeping your eyes open and collecting examples in the form of sketches, for possible use in your own design work. Shading is often a way in which building inhabitants can take some control over their own comfort.

COMBATING THE SUN
shade and comfort in the midday sun

You can shade a city...

In the cities of southern Spain, where summers are very hot, some respite is provided by fabric shades stretched across streets at roof level. Being fabric and translucent, these shades do not cut out all the light. Small gaps in the fabric also allow some glints of sunlight to enliven the scene. Slight breezes gently ripple the fabric making them a counterpoint to the solidity of the adjacent buildings. Trees also supplement the shade and increase the number of permutations for making a comfortable and attractive street.

In the southern Spanish city of Seville, where the summer sun is strong, streets are made more comfortable with fabric shades stretched across at roof level.

In Albania once I sketched a house where the owner had planted a tree to shade the front porch, clipping it into the shape of a flat umbrella (left). The result was an attractive entrance transition that was also a comfortable place to sit in green dappled shade.

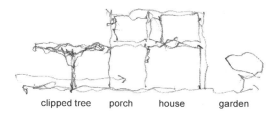

clipped tree   porch   house   garden

GETTING OUT OF THE SUN
simple shadow

… or just yourself.

Seeking shade is a timeless and universal response to exposure to the heat of the sun. We might find shade under a tree or in a cave but we can make it for ourselves too, using materials readily available.

Aboriginal Australians create shade humpies or wurleys out of twigs and scrub.

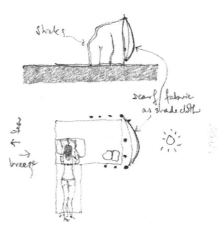

I might shade myself on the beach by pulling a towel over my head (above). Someone else might create a more elaborate shade structure using sticks and fabric (right). On page 65 there is drawing of a simple yet ingenious structure that exploits breeze to make shade with light fabric.

Provision for shade has been built into architecture since earliest times. The megarons of Mycenaean civilisation (1600–1100 BCE) were the halls of the kings (right). Homer (in The Odyssey) describes discussions taking place and guests being lodged in the ventilated shade of their porches (a).

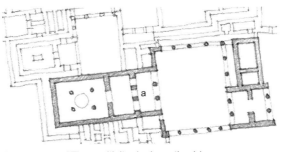

the megaron of Tiryns with its shady portico (a)

THE SUN MOVES
awnings and sun shades

We use shadow to reduce the need for energy-hungry air conditioning.

The position of the sun in the sky is variable. The shadows it casts vary too, as do their effect on the interiors of buildings. The sun moves across the sky from east to west. During the passage of the day its angle and direction change, being low at dawn and dusk and high at midday. In northern latitudes its angle is generally low, but varies from season to season. In equatorial regions it can, at noon, be right overhead. In temperate regions the sun does not reach such challengingly high angles. With all these dimensions of variation it is important that admission or exclusion of sunshine is carefully considered.

Sunshades can lessen the amount of sun entering a room and help reduce the need for artificial cooling. (See also page 170.)

Cultural attitudes to sun and shadow can be very different in different regions of the world (see page 51). In Australia, where the sun is high and hot, it can be desirable to exclude the sun completely from the interiors of buildings at all times of day. In northern countries (such as Wales) the sun, and the warmth and light it brings, is welcome. Rooms facing north, away from the sun during the day, are mostly, for the greater part of the year (except in the early mornings and late evenings of summer), devoid of the sun and its shadows. But even so, such rooms can be good studios for artists because of their even – sunless and shadowless – light.

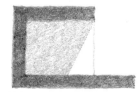
high sun, midday

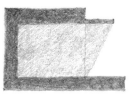
with sunshade

medium sun

with sunshade

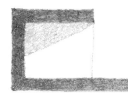
low sun

with sunshade

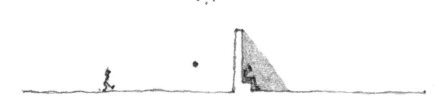

'The wall is the partner that returns the ball to the child and gives shadow to the elderly. Both sense the nearness of the elements of architecture.'

Sverre Fehn in Per Olaf Fjeld
– Sverre Fehn: The Pattern of Thoughts, 2009.

SHADING WHAT'S BELOW
in open country, by the beach, in the city...

section

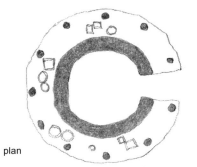

plan

traditional house, Zimbabwe

Under its thick thatched roof the circular interior of this traditional village house in Zimbabwe is contained shadow. But the eaves of the roof overhang make a shadow container, keeping the walls from being overheated and shading a low platform on which things may be stored out of the sun.

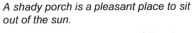

Lake Hotel, Llangamarch Wells, Wales

A shady porch is a pleasant place to sit out of the sun.

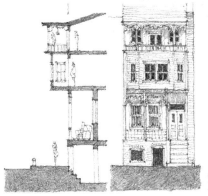

traditional timber house, Istanbul

As well as increasing the amount of living space available, the jettied storeys of a traditional Ottoman timber house in Istanbul (above) provide progressive layers of shade for the storeys below. The house has a veranda (balcony, loggia, porch... whatever you wish to call it) which is a comfortable shady place where the residents can, whilst being raised above pavement level, sit and retain contact with the life of the narrow street outside.

beach café, Mallorca

Across the world, cafés adjacent to beaches have sitting areas open to the sea breezes but shaded from the sun by light frameworks of timber and fabric.

SHADOW

A SHADY PLACE TO SIT
and watch the world

Monemvasia, Greece

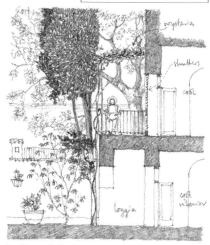

Positano, Italy

We enjoy watching the world from the psychological cover of shadow. It may be just the shade provided by the relaxed architecture of a place: the shady terrace of a restaurant or hotel; a tree in a town square; a single small, densely foliaged, tree with a panoramic view of a beautiful sunlit bay (below). In all these instances it is the shadow – and the psychological as well as physical comfort it offers – that we seek. As architects we have the opportunity to provide such enjoyable places for others.

Above is the terrace of a hotel overlooking the Italian resort of Positano. Shaded by wisteria and various trees it nevertheless has a view across the Amalfi coast. A good place for breakfast.

the shaded tree place

Tenby, Wales

SHADOW AND COURTYARDS
through (Italian) history

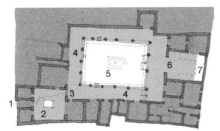

Casa degli Amorini Dorati, Pompeii

triclinium

The typical Pompeian house is an orchestration of light and shadow. Often these houses are entered on axis, but this one has a corner entrance (bottom left in the plan above). After an inset porch (1), which is a shadow threshold, it leads into a small atrium (2) lit by an opening in the sky. Another shadow threshold (3) leads through into the shady peristyle (4) running all the way around the sunny garden (5). On axis and at the end of the garden is the reception room – the triclinium (6) – which, as it stands, creates its own habitable shadow container with light at its rear (7; top right).

A typical Pompeian villa is an instrument of shade, tuned to provide environmental comfort and aesthetic pleasure.

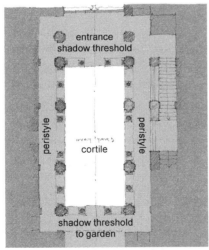

Palazzo Tamborino Cezzi, Lecce, Italy

The Italian Renaissance palazzo also has an entrance courtyard – cortile – for light and air, surrounded by a shady arcade (peristyle). In this instance the courtyard leads through to a bright sunny garden.

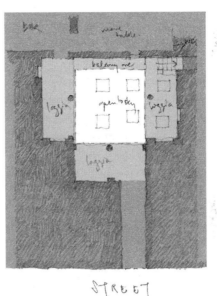

café, Orta S. Giulio, Italy

This café is set in a small courtyard off a narrow street in a small north Italian town on the banks of Lake Orta. The passageway from the street creates an attenuated shadow threshold before the relative brightness of the courtyard. The courtyard is lined by shady loggias set under the rooms or balconies of apartments above (which are also accessed from the courtyard, giving it an extra layer of vitality).

SHADOW AND COURTYARDS
providing a variety of types of shade

In your own design, experiment with providing different types of shadow.

The practical contributions shadow makes to the places we occupy can be subtle and many-layered. Below are two examples from Malta, a warm and sunny island in the Mediterranean where shade is an important necessity for summertime comfort. The upper drawing is of a courtyard tea garden set just inside the thick defensive walls of the old city of Mdina, Malta's one-time capital. The lower drawing is of a seventeenth century palazzo, now a hotel, also just inside the city's walls. Both enjoy panoramic views across the island towards its north coast. Both therefore benefit from being positioned on the cooler northerly side of the city. Interpreting my on-site sketches of their cross-sections you can see that these places provide different kinds of shade. The variety contributes to their vitality and aesthetic appeal.

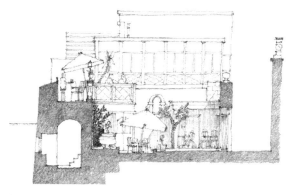

Fontanella Tea Garden, Mdina, Malta

Shade in this tea garden and the street outside is provided partly by the walls. This is supplemented by vegetation and parasols. There is also a shady loggia (towards the back in the drawing) which can be opened to island breezes. Cooling in the courtyard is also helped by the evaporation of water from a fountain (after which the tea garden is named).

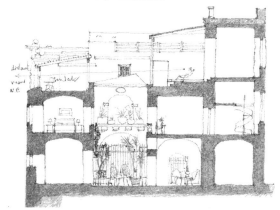

Xara Palace Hotel, Mdina, Malta

The levels of this palazzo step down from south to north across the section, so the northerly spaces, with the views, are shaded by the higher walls to the south. It has a small shady courtyard, latterly fitted with a glass roof, which would originally have been open to the sky. On the breakfast terrace at roof level there are canopies, parasols, pergolas… to provide shade with a variety of different characteristics.

PROBLEMATIC SHADOW

We see light as positive and darkness as negative. Shadows can be problematic. To be 'in the dark' about something is negative; to 'see the light' positive. To be described as being 'in someone's shadow' – metaphorically as well as actually – is seen as a sleight; 'hogging the limelight' may be arrogant but it gets attention. 'Lurking in the shadows' is suspect; 'in the full light of day' revealing.

Most of this Notebook is about the positive possibilities of shadows in architecture. In a world keen to optimise illumination and flood spaces with an even light (see the Ingold quotation on page 10) – mainly for functional reasons but also as a subliminal ideological urge for openness, visibility, accountability – the positive and poetic possibilities of shadows have been neglected. The redress of shadow is due.

Nevertheless, shadows can cause problems, especially when you cannot quite see something that you want or need to; or when you would like to enjoy some sunlight but something is in the way. And as architects we owe neighbours some consideration, and should try not to put them in the shade.

PROBLEMATIC SHADOW
writing and reading

Writing and reading are both easier if you are not in your own light.

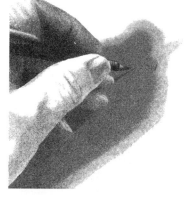
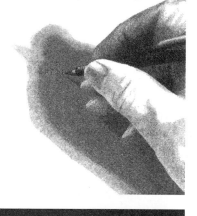

left: 'A Man seated at a table in a lofty room', c.
1628–30, sometimes attributed to Rembrandt
above: another man reading in the shadows

PROBLEMATIC SHADOW
cricket and 'right to light'

Shadow can interfere, discomfit, spoil.

Late in a day's play, shadows can cause problems in cricket. A fielder might want to stand close to the batsman to optimise chances of a catch.

But if the fielder's shadow is cast in the area where the ball will bounce (above left) they will have to move back (above right), not to distract the batsman.

Interview panels often sit with their backs to a window. Their argument for doing this is that it means that the light from the window falls on you (the interviewee) and that they can see you more clearly. A consequence is that, from your point of view, their faces are in shadow and it is more difficult for you to read the nuances of their expressions. Sometimes one suspects they might do this on purpose to seem more sinister! In which case it is a conscious attempt to exploit a problematic shadow by means of social geometry.

One of the reasons we go on holiday to the beach is to enjoy the sun. But in some places, particularly on east-facing beaches, the development that accommodates holiday-makers frustrates their reason for being there. On the east coast of Australia, for example, there are places where the shadows of high-rise apartments blot out the sun from most of the beach during the afternoon (above). This is an instance where architecture is in harmony with the vector (see page 15) of topography (the view across the ocean) but not of the sun.

RIGHT TO LIGHT
not putting others in the shade

In temperate climes, where you would appreciate some sun, you might regret building your house in the shade of a mountain (above) or at the bottom of a gloomy valley (right).

In some countries developers and their architects are constrained not to put neighbours in the shade. It's called a 'right to light'. In other countries such considerateness is not imposed by law. Even so, neighbourliness implies that aggravation is likely if a large development (on a beach front for example; below) both removes a view and obstructs sunshine from reaching long-established smaller properties.

SHADOW CONTAINER

One of the shortcomings of conceiving of architecture as primarily a matter of appearances (the ornamentation, proportion and modelling of surfaces) is that something more fundamental – identification of place – can fall out of focus. The subtleties of place-making can be submerged beneath a preoccupation with style, sculptural form and mechanical planning formulae. There is a lot more to shadow in architecture than graphic effect (pages 35–44). Shadow plays an important role in place-making. One might argue that architecture originates in our desire to find refuge in shadow.

'Whenever life seeks to shelter, protect, cover or hide itself, the imagination sympathizes with the being that inhabits the protected space... Shade, too, can be inhabited.'

Gaston Bachelard, trans. Jolas (1964) – *The Poetics of Space* (1958), 1969.

OUR EARLIEST ARCHITECTURE
shadow of tree or cliff

Choosing shadow as a place to sit is an act of architecture.

l'abri de Cro-Magnon, Les Eyzies

rock shelter, Ayer's Rock, Australia

'So the LORD God appointed a plant and it grew up over Jonah to be a shade over his head to deliver him from his discomfort. And Jonah was extremely happy about the plant.'

Jonah 4:6.

A PLACE TO LIVE
shadow container of a tree

The shadow container has a claim to be our very first work of architecture. On page 14 I mentioned the shadow container provided by a rock overhang as a place to settle out of hot sun (opposite top, in France and Australia). Trees too provide refuge (opposite bottom). There is something natural about sitting in the shade of a tree. All animals (including ourselves) seem to do it instinctively (right). In Mediterranean and other hot climates the shade of a tree can become a living room (below).

The shadow of a tree is a living room.

SHADOW CONTAINER
defined architecturally

Trees' shadows define their own area of ground. Their edge is the threshold one crosses to enter their shadow container. Sometimes that edge can be reinforced into a boundary barrier by means of a fence or wall. Once in Albania I saw a pen positioned under a tree to contain sheep in its shadow during the hottest part of the day (right).

Taking his influence from village practice in Africa, the architect Francis Kéré designed a pavilion for London's Serpentine Gallery inspired by the idea of shade under a tree. He too defined and protected the area of shade with a wall (below). The 'tree' was a stylised canopy which filtered the light from the sky.

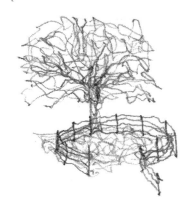

In this simple work of architecture the edge of a shadow threshold has been outlined with a fence – a pen for sheep.

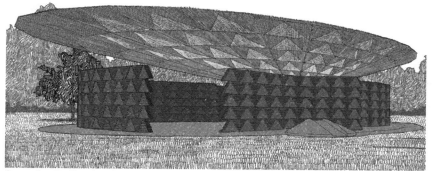

Serpentine Pavilion, London, Francis Kéré, 2017

'What I like so much about the form of a tree is that it creates a shelter – shade to protect from the sun, or cover from the rain. It also serves as a meeting point for a community, it's not closed to the outside world. You can relate that directly to architecture: first you need a big canopy to protect you from the sun. Then you need walls to protect you against wind and rain, but also to create a sort of intimacy, to embrace you. The space between two trees shares that kind of community tradition. In some cultures and climates, during the day you spend more time under a tree than inside a house.'

Francis Kéré, interviewed by Fiona Shipwright, in 'Of Clay and Community', *Mono.Kultur #46*, Berlin, Autumn 2018.

SITTING IN THE SHADE
shade and community

Shadow holds people together.

Bedouins sit together in the shade of a tent.

Since ancient times one of the primary purposes for rudimentary architectural construction has been the creation of shade. This may be through choice or necessity. Shade can be the bonding medium that holds people together.

Syrian refugees make themselves some rudimentary shade.

'Shadow, let not men perish through the burning heat.'

Sir E.A. Wallis Budge – *The Literature of the Ancient Egyptians*, 1914.

SHADOW CONTAINER
Keralan mud and thatch house

I included this Keralan mud house as Case Study 4 in *Analysing Architecture* (4th ed., 2014). I neglected to include the illustrations below, which illustrate how this simple but subtle dwelling is primarily a shadow container – created by its thatch roof – for the life lived within it.

The eaves are very low – you have to bend to go under them – so the interior of the house is always in shadow.

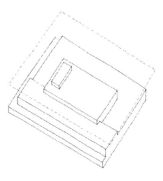

The primary compositional elements of the house are a stepped platform under a thatched roof.

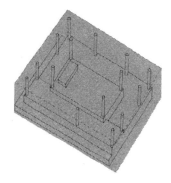

The roof creates a shadow container that embraces all the house's living spaces defined by the different levels of platform.

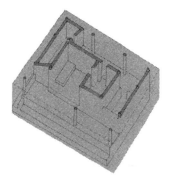

Those living spaces are further defined by a few mud walls that only completely enclose places where valuables are kept.

The result is a house that is given integrity by its accommodating shadow, which makes it a comfortable place.

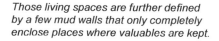

SHADOW CONTAINER
Robie House, Chicago, Frank Lloyd Wright

Frank Lloyd Wright designed a number of so-called 'Prairie Houses' during the first decade of the twentieth century. Their appearance in a portfolio of his work published in Germany in 1910 influenced the European development of Modern architecture. The most distinguished of the Prairie Houses is the Robie House in Chicago. Its design is characterised by a clear distinction of the elements of wall and roof, and by the strong horizontal emphasis given to the whole composition. The deep overhanging eaves of the low-pitched roofs create shadow containers shrouding the living spaces in privacy (top right). Some fifty years later, setting down 'Principles' for 'The New Architecture', Wright identified shade as an essential component in lending 'charm', 'style' and 'significance' to architecture.

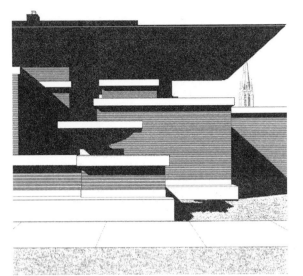

Robie House

The deep eaves of the Robie House are an important component of the horizontal emphasis that Wright gave his design for the house. That emphasis was thought to harmonise with and evoke the wide horizons of the American prairie. Those deep-eaved roofs also create shadow containers accommodating the living spaces and enhancing their privacy. Throughout his career Wright considered shelter and its corollary shade as essential elements in the production of domestic architecture. (See the quotation below, written half a century after the design of the 'Prairie Houses'.)

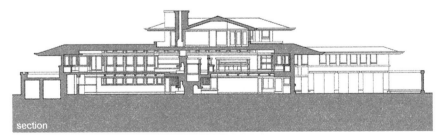

section

'Almost all... features of design tend to lead by one another to this important feature, shelter, and its component shade... The occupants of a building readily discover greater opportunity for comfort and more gracious, expanded living wherever shelter is becoming shade. By shade, charm has been added to character; style to comfort; significance to form.'

Frank Lloyd Wright – 'The New Architecture: Principles' (1957), in Kaufmann and Raeburn, eds. – *Frank Lloyd Wright: Writings and Buildings*, 1960.

SHADE FOR ROYALTY
Knossos and Barcelona Pavilion

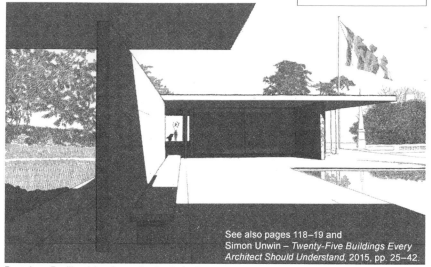

See also pages 118–19 and Simon Unwin – *Twenty-Five Buildings Every Architect Should Understand*, 2015, pp. 25–42.

Barcelona Pavilion (view from c in plan below)

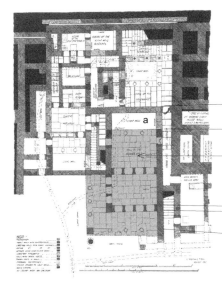

Correspondences suggest that Mies van der Rohe may have been influenced by the Knossos plan (left) when designing the Barcelona Pavilion. It too would house a 'throne room', for the King and Queen of Spain. It is also a shaded place, complete with comparable light well (b).

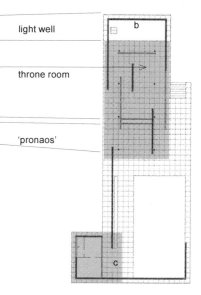

light well

throne room

'pronaos'

The plan above was published in the magnum opus of Sir Arthur Evans, archaeologist of Knossos, The Palace of Minos, *in 1921. It shows the Hall of the Double Axes, which he interpreted as a throne room. Responding to the hot Cretan climate, this is a shaded space, the light and cross-ventilation of which is helped by a distinctive light well (at a).*

plan of Barcelona Pavilion (1929)

SITTING IN THE SHADE
at a bullfight

Shadow can cost more.

Bullfights in Spain are fought in great circular amphitheatres.

The bullring (above) is an architectural lens that focuses the attention of the spectators on the drama of the fatal action. It takes the classic form of the circle to encompass a ceremony of death in the afternoon.

The drama is compounded by the strong Spanish sunlight. Sometimes the circle of death is in bright sun.

But because of the asymmetrical path of the sun some parts of the spectator seating (to the south and west) are in shade. The shaded seats are more expensive. Few have the stamina to last an afternoon of killing exposed to the full strength of the sun. But perhaps the cast shadow has a poetic purpose... it veils the watchers, the witnesses of slaughter.

'The west walls of the bull ring building cast a shadow and those seats that are in the shade when the fight commences are called seats of the sombra or shade. Seats that are in the sun when the fight commences but that will be in the shadow as the afternoon advances are called sol y sombra. Seats are priced according to their desirability and whether they are shaded or not. The cheapest seats are those which are nearest the roof on the far sunny side and have no shade at all at any time. They are the andanadas del sol and on a hot day, close under the roof, they must reach temperatures that are unbelievable in a city like Valencia where it can be 104° fahrenheit in the shade, but the better seats of the sol are good ones to buy on a cloudy day or in cold weather.'

Ernest Hemingway – *Death in the Afternoon*, 1932.

Under the fierce Spanish sun the most expensive seats for bullfights are those in shadow.

URBAN SHADE
Parasol, Seville, Jürgen Mayer

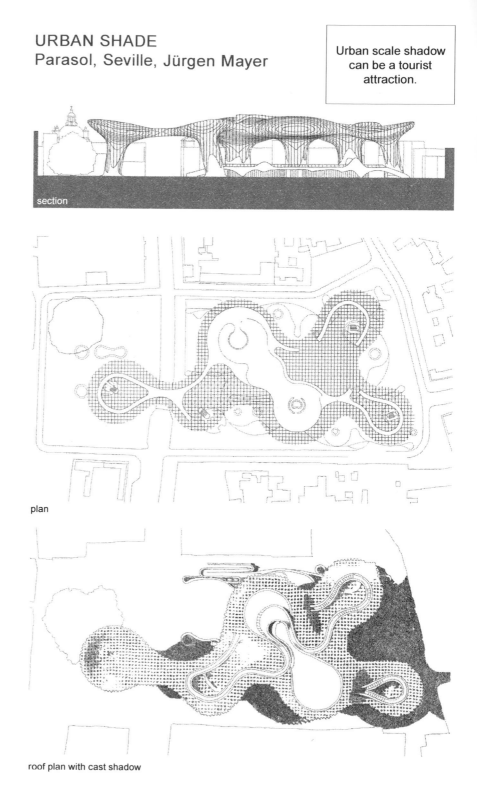

section

plan

roof plan with cast shadow

URBAN SHADE
building clouds under a hot Spanish sun

All architecture with roofs and walls casts a shadow. But it is rare for the primary and dominating purpose of a work of architecture (apart from temporary examples like umbrellas, canopies, awnings…) to be to cast a shadow. But this is the case with the Metropol Parasol, designed by German architect Jürgen Mayer and completed in 2011. It stands in the Spanish city of Seville which suffers strong sun in the summer.

If the function of the Parasol is to help make an urban square more comfortable by obstructing solar radiation then its other architectural purpose is to act as an attention grabbing ornament for the city, an item of 'urban bling' that will attract photographers and their editors to advertise the city across the world. It is also a 'climbing frame' with visitors allowed to ascend to a roof walk.

The primary function of the Metropol Parasol in Seville is to provide an urban square with shade, so as to make it a more comfortable place to be in the summer. But the Parasol is more than merely pragmatic: it is sculptural, and an advertisement for the city.

SHADOW UNDER THE HOUSE
traditional Queensland houses

The shade under a house is a comfortable place out of the sun.

In Queensland, Australia, houses are built on timber posts (stumps) to increase ventilation and to lift them clear of snakes and termites. Doing so also creates a shady place underneath the house. On hot summer days this is the most comfortable place to live. (With their metal roofs, the rooms inside the house can heat up like an oven.)

This traditional Queensland house is lifted above uneven ground on stumps. The shadow of the house provides a place of respite from the hot sun.

In temperatures over 30°C (86°F) the most comfortable place to sit is in the shade under the house.

(The lower drawing is of our house when we lived in Queensland in the early 1980s.)

SHADOW UNDER THE HOUSE
Glass House, São Paolo, Brazil, Lina Bo Bardi

In her own house (1949–52), Lina Bo Bardi did something similar. The large overhang, supported on metal posts, creates a shady place under the house.

The overhang also creates a shadow threshold for the house.

See also Simon Unwin – *Twenty-Five Buildings Every Architect Should Understand*, 2015, pp. 223–32.

SHADE ROOF
Minoan house, Crete, c. 1700 BCE

Shade has always been an essential commodity in hot climates. As well as under houses, shady terraces can be made on their roofs, where there might also be some breeze.

Above is a clay model of a house with a shaded roof terrace. It dates from around 1700 BCE and is exhibited in the Heraklion museum on Crete. The island was inhabited around three-and-a-half thousand years ago by the Minoan civilisation also responsible for the great palaces of Knossos and Phaistos.

The clay model Minoan house has been provided with a plausible modern replacement for the original shade roof it clearly had; though instead of the canopy shown it may have been shaded by vegetation supported on a trellis. You can imagine this shaded roof terrace as a congenial and comfortable place for a warm Cretan afternoon and evening.

The sophistication of house design in ancient times can be greater than we might think. This nearly four thousand year old model of a Minoan house is clearly based on real houses. Maybe it was a play thing for a child. Even so, it represents a climatically sensitive strategy for a comfortably shady roof terrace, which would also catch whatever breeze might be going, for comfort during hot Cretan summers.

SHADE ROOF
Roof-Roof House, Ken Yeang

Ken Yeang used a similar strategy in his Roof-Roof
House in a suburb of Kuala Lumpur (1964), here
recorded and analysed in one of my notebooks.
It, like the Minoan house, has a canopy shading
the house and its roof terraces from the Malaysian
sun. This canopy is louvred to provide different
levels of shade at different times of day.

*In the Roof-Roof House
the canopy helps reduce
the energy needed for
cooling the house, by
shading it from the sun.*

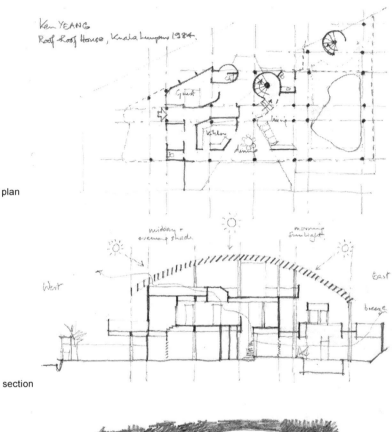

plan

section

pictorial view

SHADOW OF DEATH
eternal shade

Tombs are shadow containers.

Monument to Mrs Soane, John Soane, 1816 (drawn by J.M. Gandy)

In 1816, John Soane designed a monument to his
wife (above). It became the basis for his own much
smaller tomb, which stands in St Pancras Gardens
(the graveyard of St Giles in the Fields) in London. The
design shelters the sarcophagus under a canopy, in the
shadow container of death.
　　　　When Giuseppe Brion died in 1968 his wife
Onorina commissioned Carlo Scarpa to design a private
burial ground adjacent to the cemetery in San Vito
d'Altivole in Italy. He designed twin sarcophagi for them,
leaning in towards each other, sheltered in the shade
of a bridge-like structure (below). Like an ancient burial
mound it is oriented towards the midwinter sunset and
the midsummer sunrise.

Brion tombs, aerial view

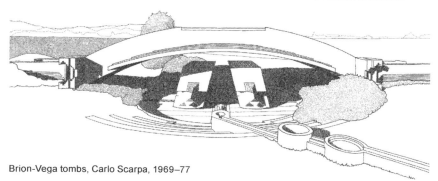

Brion-Vega tombs, Carlo Scarpa, 1969–77

CONTAINED SHADOW

Shadow contains but it can also be contained. Shadow containers can provide us with cool places out of the hot sun. A simple box or completely enclosed room contains shadow. The moment before a performance, when the lights have been extinguished, an auditorium is one large contained shadow waiting to be broken by the lights revealed when the curtain rises. The audience, shrouded in darkness, continues to sit in the refuge of its contained shadow throughout the performance, witnessing the action through a shadow frame.

The ultimate contained shadow is complete claustrophobic darkness, as in the cupboard under the stairs, a dungeon, an unlit attic, the grave...

But contained shadow can also be the precondition of more subtle effects of light and shade, effects that are not possible in a bright, evenly lit space. Shadow is sometimes described as a pool; things can swim together in its water.

DARK REFUGE
Chapel, Oberrealta, Christian Kerez

Contained shadow is
a refuge from
the world.

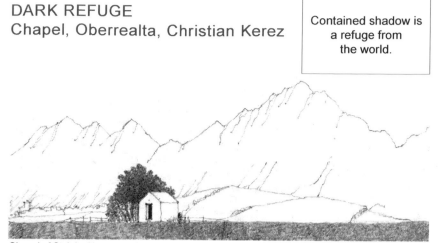

Chapel of St John Nepomuk, Oberrealta

entrance elevation

Escaping agoraphobic exposure to a
large and sublime landscape into the
refuge of a dark cell is one of the timeless
architectural experiences. Architecture
since deepest prehistory has been
motivated by this phenomenological
drive, this desire for a shady escape from
the world.

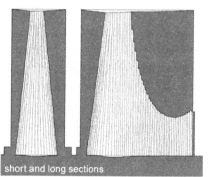

short and long sections

plan

*The Chapel of St John Nepomuk (1993;
above) is the simplest form of chapel
set on the edge of a village and with a
backdrop of Swiss hills. In the tradition
of countryside chapels it offers a small
shadowed refuge from the world. Entering
it feels like entering your own head.*

*Peter Zumthor's Bruder Klaus Chapel
stands in open fields under a big sky
some twenty miles south-west of Cologne
in Germany. It was built in 2007. Like
the Chapel of St John Nepomuk it offers
a small dark interior as a stark contrast
with its open surroundings. Here, that
darkness is intensified by the black
charring of the interior surfaces by the
burning of the logs that had made the
interior formwork for its poured concrete
structure. This gives the inside of the
chapel a particular smell that might claim
to be the perfume of shadow. Light enters
from an opening above.*

CAMERA OBSCURA
Both Nam Faileas, Chris Drury

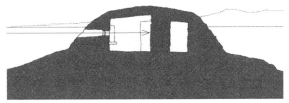

section

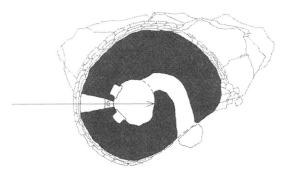

plan

Chris Drury's 'Hut of the Shadow' on the coast of the island of North Uist is an interpretation of an ancient burial chamber – with its contained deep shadow – as a camera obscura. Being in it is also like being inside a partially sighted eye with a rock retina where the reality of the world outside is shown merely as a dim shadow of itself (cf. Plato's 'Metaphor of the cave', page 13).

section

In the mid 1990s, on the coast of the island of North Uist in the Outer Hebrides, the artist Chris Drury constructed a building in stone that appears like an ancient burial mound. It has a small chamber entered by a simple labyrinth passage way (to keep out light). When you go in, the chamber seems completely dark except for a hole in the wall opposite through which enters a small amount of light. Eventually your eyes adjust to the darkness. You realise there are simple seats set into the wall. Then, as your eyes adjust further, you see that the light entering through the hole in the wall is actually projecting a view of the external seascape onto the flat wall opposite. The 'burial chamber' is a camera obscura, a 'cinema' where the 'movie' is the real world outside.

The show in this primitive cinema depends on the deep shadow contained inside the building. Without it the camera obscura would not work and the novelty of seeing the world as a ghostly moving image in the privileged location of a completely dark refuge would not be available.

Shadow is both the image and the necessary condition for its projection.

Camera obscuras always depend on contained shadow. Otherwise the faint images projected from outside could not be seen. This is a portable sedan chair camera obscura, for the use of artists, made in the eighteenth century.

For more camera obscuras see: John Hammond – *The Camera Obscura*, 1981.

PLANETARIUM
containing universal shadow

The universe is infinite shadow.

We think of the universe as infinite darkness sprinkled with twinkling stars (and the fatal vortices of black holes). When we wish to model the night sky, a precondition is darkness. To achieve this we construct containers of shadow. By means of these enclosed rooms we make architecture to accommodate the grand and infinite shadow of the whole universe.

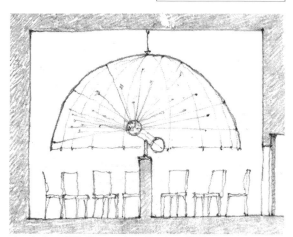

planetarium, Mills Observatory, Dundee, 1935

In Balgay Park, Dundee, there is an observatory, built for the public to view the stars through a refracting telescope. Inside is a tiny planetarium (above), a miniature model of that infinite universe seen through the telescope. Its form is a large hemispherical black umbrella situated in a dark room. The stars are projected onto the under surface of the umbrella.

A much grander example of architecture constructing a container for the infinite universal shadow is the cenotaph (empty tomb) that the French neoclassical architect Étienne-Louis Boullée designed for Sir Isaac Newton (below). Never built, it would have been a huge dome – a complete sphere of dark contained shadow – perforated by stars, with Newton's memorial at its focus.

Cenotaph for Sir Isaac Newton, Étienne-Louis Boullée, 1784

ANALYSING ARCHITECTURE NOTEBOOKS

COALESCED BY SHADOW
John Soane's Museum

Contained shadow is like a deep pool of water.

Shadow holds things together in collective gloom.

John Soane Museum, London

John Soane added the museum to the back of his house in Lincoln's Inn Fields in 1809. It contains a multitude of generally monochrome exhibits. Rather than each being given its own pedestal and spotlight – as in modern museums – they are all submerged in common shade, dimly illuminated by roof-lights. The museum has various levels, with the most profound shade at its deepest. Like water, the contained shadow binds all together as a complex but unified community of objects.

Even though Soane sought 'lumière mystérieuse'*, what holds all these objects together is not so much the light but the shade in which they are collectively immersed.

* John Soane – *Lectures on Architecture* (Lecture 8), 1929.

SHADOW IN THE TEMPLE
Lincoln Memorial, Henry Bacon

A father watches over his nation from contained shadow.

Shadow contributes to the poetry of architecture. The Lincoln Memorial in Washington (1922) is a case in point. Above is the front (east) elevation depicted as an architectural drawing without shadows. Below is the same view during the latter part of the morning, with the sun in the south-east.

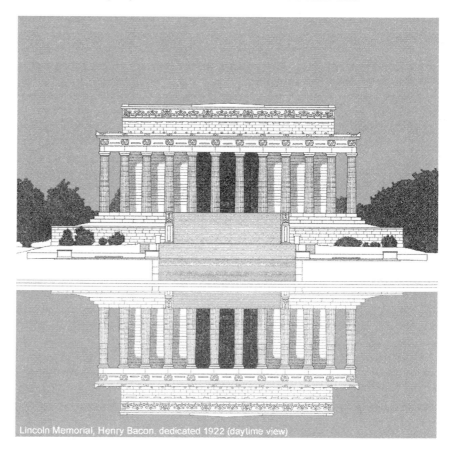

Lincoln Memorial, Henry Bacon, dedicated 1922 (daytime view)

SHADOW IN THE TEMPLE
contained shadow as shrine

Like most architecture the Lincoln Memorial is something of a solar clock. In the drawing opposite you can see the shadows of the colonnade on the wall of the cella (inner chamber) behind. As in the elaborate Beaux-Arts renderings produced by architects in Bacon's time (see pages 35–42), those shadows, because of the direction of the sun, provide an asymmetric counterpoint to the axial symmetry of the memorial's architectural composition. They also indicate the depth of the peristyle of columns around the cella giving the elevation, in reality as well as in drawing, its three-dimensional depth.

But poetically the most powerful shadow in the composition is the one that fills the central opening of the cella, on the axis of which sits the giant statue of Abraham Lincoln himself. The gloom of this contained shadow shrouds Lincoln in the shadow of death, from the cover of which he continues to watch over the nation.

At night (below left), the floodlit statue of the dead president becomes more visible within the illuminated casket of the memorial. The following morning, at dawn, the sunlight will strike into the temple to light it up in a different way, as a moment of resurrection.

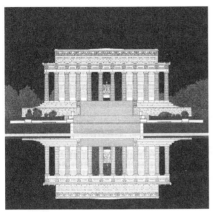

Lincoln Memorial (night view)

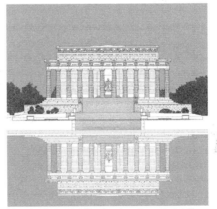

Lincoln Memorial ('by the dawn's early light')

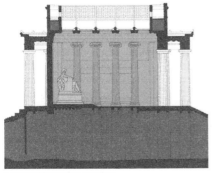

Lincoln Memorial, section

The interior of the memorial is not in the deepest shadow. It is infused with the warm glow of light filtering through a translucent marble ceiling from rooflights above (see the section left). The shadow in which the statue of Lincoln is immersed is not the negative shadow of light completely withheld; it is shadow poetically modified, as if to suggest that death (at least for the principles of someone like Lincoln) is not total eternal dark nothingness.

PRIMITIVE GLOOM
St Petri Church, Sigurd Lewerentz

> Shadow is the realm of primeval mystery, religious and secular.

In *The Stones of Venice* (1851–3), John Ruskin describes entering St Mark's Basilica in Venice:

> 'Through the heavy door whose bronze network closes the place of his rest, let us enter the church itself. It is lost in still deeper twilight, to which the eye must be accustomed for some moments before the form of the building can be traced; and then there opens before us a vast cave, hewn out into the form of a cross, and divided into shadowy aisles by many pillars. Round the domes of its roof the light enters only through narrow apertures like large stars; and here and there a ray or two from some far-away casement wanders into the darkness, and casts a narrow phosphoric stream upon the waves of marble that heave and fall in a thousand colours along the floor. What else there is of light is from torches, or silver lamps, burning ceaselessly in the recesses of the chapels: the roof sheeted with gold, and the polished walls covered with alabaster, give back at every curve and angle some feeble gleaming to the flames; and the glories round the heads of the sculptured saints flash out upon us as we pass them, and sink again into the gloom.'

The church, like many others, is shadow contained. Ruskin's description could have been part of the inspiration for Lewerentz's Church of St Petri in Klippan (1963–6).

The interior of St Petri, as in Ruskin's description of St Mark's, is a cave of shadows (though square rather than cruciform). Like in St Mark's, the floor of St Petri (though brick rather than marble) is uneven. As in St Mark's, there are places in St Petri where 'light enters only through narrow apertures' in the roof, and where 'a ray or two... wanders into the darkness, and casts a narrow phosphoric stream' upon the waves of the floor; 'what else there is of light is from... lamps, burning ceaselessly in the recesses'. Unlike in St Mark's, Lewerentz chose to give his church generous square plain glass windows, the glare from which intensifies the dark extremities of its contained shadow (see page 105).

See also *Twenty-Five Buildings Every Architect Should Understand*, 2015, pp. 175–86.

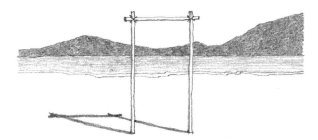

SHADOW THRESHOLD

In the quotation below, Joseph Conrad uses 'shadow-line' as a metaphor for a rite of passage, the transition from youth to maturity. In architecture transition means movement from one place into another, usually across some sort of threshold. Such movement, such transition, often involves shadow. It may even sometimes be the case that the transition only involves moving into, out of, or across a shadow-line. Experiencing such shadow thresholds may be noticed only subliminally, but they are nevertheless significant and always affect us in some way, however slight and imperceptible that effect might be. The basic, and timeless, shadow transition is the movement from the bright outside into the mysterious shade inside of a cave... or vice versa (an echo of birth). As well as in rites of passage, shadow thresholds play a part in religious or aristocratic architecture, often at an entrance where one passes from the ordinary world into a sacred place or one that its owner wants to present as specially privileged.

'It seems to me that all my life before that
momentous day is infinitely remote, a fading
memory of light-hearted youth, something on   Joseph Conrad – *The*
the other side of a shadow.'   *Shadow-Line* (1916).

SHADOW TRANSFORMS
healing... cursing?

Passing through a shadow can change you, for better or worse.

To imbue shadow with powers of transformation is not the same as using the idea of shadow as a metaphor. In the latter, the idea of shadow is transferred to ideas that might not involve light and its absence. But the former suggests that we can interpret actual shadows as possessing some power over us and our actual – physical or psychological – state of being.

A commentary by nineteenth-century Scottish theologian Dr Hugh MacMillan includes some observations on the beneficial powers of shadows. He does so metaphorically but also by allusion to actual shadows:

> 'If they place themselves in their simple faith under the shadow of God's house, the blessing will assuredly not be wanting... The shadow of a tree or rock is a very delightful and refreshing thing on a burning summer day. It cools the heated frame, and imparts vigour and strength to the languid body.'*

He was commenting on the following passage from the New Testament, which tells of the faith some had in the potential healing power of the shadow of St Peter.

> 'Insomuch that they brought forth the sick into the streets, and laid them on beds and couches, that at the least the shadow of Peter passing by might overshadow some of them. There came also a multitude out of the cities round about unto Jerusalem, bringing sick folks, and them which were vexed with unclean spirits: and they were healed every one.'**

The story was illustrated by Masaccio in a fresco (top right) in the Brancacci Chapel in Florence.

'St Peter Healing the Sick with His Shadow', Masaccio, c. 1425

We can see shadow thresholds as maleficial too. Our reluctance to walk under a ladder propped across a pavement may suggest we can ascribe negative, unlucky, effects to passing through shadows. Shadow is a common metaphor for the threshold of death. Passing through a doorway into a deeply shadowed and unknown interior is always going to incite some trepidation.

But shadow thresholds can also be harbingers of better things, or places. The shadow of the gate into a sunny walled garden (see page 28) or the green shade of a Paradise courtyard serves only to intensify the pleasant effect of the place we reach once we have passed through.

* H. Macmillan LL.D., available at: biblehub. com/sermons/auth/macmillan/the_healing_ shadow.htm (September 2019).
** Acts 5: 15–16.

SHADOW AND EMOTION
into the dark, out to the light

Shadow thresholds
prompt fear
and thrill.

Even when not directly associated with life, death and retribution, shadow thresholds affect our emotions.

In the *Children as Place-Makers Notebook* (pages 28–9), I alluded to the emotional effect of crossing thresholds from light to dark and vice versa in a child's experience of the world (right). The psychological impact of such crossings stays with us through our lives. Shadow adds a nuance to the profound potential thresholds have in punctuating our general experience of the world;* and so offers a power that we, as architects, can use intentionally to orchestrate people's emotional experience of spaces and places.

It should be restated that shadow thresholds can work both ways. And that they can be combined to provide a shadowy interlude between two light places, or even a light interlude between two places of shadow. Other permutations are of course possible. The power of the shadow threshold is to add drama as well as transition.

* See *Doorway*, 2007.

As children (and adults) we experience trepidation when we enter a strange dark place... Or perhaps we feel relief at entering a refuge that we know, escaping the public gaze...

We can also feel a thrill at leaving our shady refuge to explore the bright world outside... Or we might worry that we are leaving a place of protection where we felt safe.

SHADOW THRESHOLD OF DEATH
gates of farewell

The end of life is a shadow threshold.

lychgate

As do all thresholds, shadow thresholds mark transitions between one place, one state of being, and another. Given the shadow's metaphorical association with death, it is understandable that real shadows, created architecturally, can be symbolically associated with death, whether it occurs naturally, through illness or old age, or as a matter of retribution.

Contained shadows (as well as shadow containers) have their thresholds. In Aeschylus's play Agamemnon, *all the murders (executions) take place beyond the shadow threshold of the ever-present doorway – an emblem of the ever-present prospect of death – inside the contained shadow of the cursed palace.**

* See pages 136–7 of *The Ten Most Influential Buildings in History: Architecture's Archetypes,* 2017.

Coffin bearers pause with the coffin under the shade of the lychgate before proceeding to the church for the funeral service. That shadow threshold marks the transition from the dead person's life on earth into whatever lies beyond death.

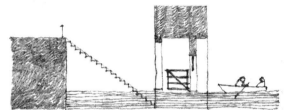

Traitors Gate, Tower of London

On their way to meet their fate, traitors would pass under the shadow of their eponymous gate as they entered the Tower of London from the River Thames.

palace of *Agamemnon*, Aeschylus (5th century BCE)

SHADOW ENTRANCE
celebrating entrance... or exit?

In his landscape intervention 'Design of a door to enter into darkness'* (above), Ettore Sottsass uses simple architecture – a rudimentary doorway constructed of poles and large leaves – to identify a place – the threshold between sunlight and the deep shadow cast across a valley by a mountain – and give it poetic significance. By doing so Sottsass invites experience of that threshold, which we might otherwise not notice. He also imbues it with allegorical power. By celebrating this threshold with his simple doorway he recognises the poetry of every passage we make between light and shadow, and the way those transitions from light to dark resonate with and presage the final passage we must all expect, from life into death. As such the simple doorway becomes a shrine and a poem.

Sottsass's 'Door to enter into darkness' is an instance of architecture becoming poetry. It combines human agency – the rudimentary construction – with natural providence – the grand shadow whose threshold the doorway marks as an invitation to enter.

* See Ettore Sottsass – *Design Metaphors*, 1988, pp. 16–17.

SHADOW THRESHOLD
Woodland Crematorium, Stockholm, Gunnar Asplund

The buildings and landscape of Skogskyrkogården (the Woodland Crematorium on the outskirts of Stockholm) were designed by Gunnar Asplund and his then architectural partner Sigurd Lewerentz between 1915 and 1940. Both architects had a keen understanding of the poetic potential of architecture. Shadow thresholds were part of their architectural syntax.

Asplund designed the Woodland Chapel (on this page) in 1918. Lewerentz completed the Chapel of Resurrection (opposite page top) in 1925. And Asplund completed the main crematorium building in 1940.

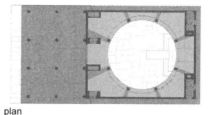

section

plan

The Woodland Chapel is approached through the dappled shade of pine woods (right). Ahead is the deeper shade of the chapel's generous porch. This forms the shadow threshold through which mourners pass to enter the light interior (below right). There the coffin lies on a catafalque under the bright sky-like domed ceiling, as in a clearing fringed by the shade of woodland – here formalised as classical columns.

Woodland Chapel, approach

'The idea of a ritual sequence – from dark to light, from sorrow to expiation, from the fear of death to the promise of life everlasting – pervades both the cemetery landscape and its architecture.'

Caroline Constant – *The Woodland Cemetery: Toward a Spiritual Landscape*, 1994.

Woodland Chapel, interior

PATH OF TEARS
Way of the Seven Wells

Way of the Seven Wells

The Way of the Seven Wells is a pathway of mourning, deeply shaded by the forests either side, leading to the Chapel of Resurrection. It too has a porch creating a shadow threshold which mourners must cross to enter a brighter interior.

approach, looking north, evening

approach, looking south, morning

In the evening the prominent Cross at the centre of the landscape casts a shadow threshold across the path to the Main Crematorium. In the morning trees cast their ghostly shadows across the path.

Chapel of Resurrection

Way of the Seven Wells

Woodland Chapel

Main Crematorium

Cross

approach

N

site plan

The buildings of the Woodland Crematorium are spread amongst a modern ritual landscape, evocative of those from prehistory. The approaches to all the chapels are imbued with poetry.

SHADOW THRESHOLD
Zumthor's phenomenology

A shadow threshold
separates the special
from the everyday.

section

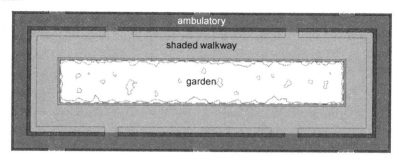

ambulatory

shaded walkway

garden

Peter Zumthor has used shadow thresholds in some of his works. He uses them as separators between places and to suggest dislocation in historical time.

The entrance to the Archaeological Shelter for Roman Ruins in Chur, Switzerland (1986), is a sheet steel box-like stepped tunnel (below, and section, right) that provides a shadow threshold dividing your experience of the remains of the past from the present world outside.

The dark ambulatory around the garden of his 2011 Serpentine Pavilion (above) was an essential introduction to the colourful sunlit garden in its courtyard. Its isolation from its surroundings was intensified by the phenomenological experience of the dark passage.

archaeology shelter entrance

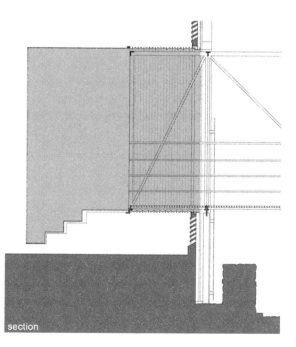

section

ANALYSING ARCHITECTURE NOTEBOOKS

NARRATIVE SHADOW

When drawing buildings that have not yet been built it is easy to take for granted that they will cast shadows. But buildings' shadows can play a significant part in the ways in which they are experienced. Opposites complement each other. Shade enhances appreciation of sunlight and vice versa. A combination of both can be more stimulating than the blinding monotony of either on its own. A searchlight focuses attention on a particular place or object while all else is shrouded in shadow. A pathway can lead in and out of shadow, playing with your emotions and perceptions along the way. Orchestrating shadow and its relationship with light is one of the abiding challenges of architecture. It can constitute the music and the poetry of architectural experience.

'My own shadow is a marker of my mass. The shadow is a visual understanding of a mass concept.'

Sverre Fehn, recorded by Per Olaf Fjeld in, Per Olaf Fjeld – *Louis I. Kahn: Nordic Latitudes*, 2019.

LIGHT OF GOD
Capel Pen-Rhiw, St Fagans

Shadows can intensify the power of oratorical preaching.

Originally built as a barn, Capel Pen-Rhiw was made into a Unitarian chapel in the eighteenth century. Box pews were installed, along with a seating gallery on an upper level. The pulpit was placed on one of the long sides of the building, with a window behind the preacher. This would have lit his text, but it would have also increased the drama of the event. With the light behind him (metaphorically as well as actually), the preacher's form would have appeared to the congregation in shadow, as a silhouette.

Unitarian chapels make a virtue of simplicity.

section

'For we are sojourners before You... as all our fathers were; our days on the earth are like a shadow, and there is no hope.'   1 Chronicles 29:15.

The chapel is simple in form but not without subtleties. The pulpit, at a, stands near the centre of one of the long walls. From here the preacher, dramatically shadowed against the light, commands the attention of the whole congregation. Silhouetted, the source of his sermon becomes otherworldly.

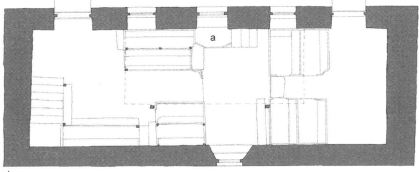

plan

A SUNLIT PROSPECT
morning shadow

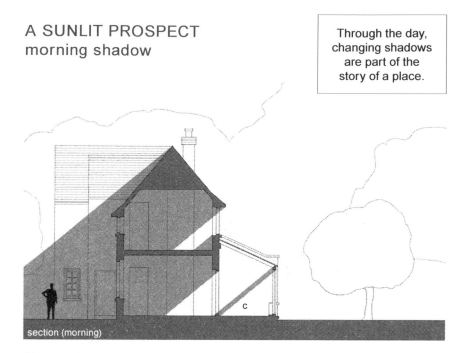

section (morning)

Shadow plays its part in our experience of more ordinary dwellings too. In cooler northern climates we can appreciate a room that is warmed by the sun. In hotter regions shady rooms are more comfortable.

On a sunny morning our house (above), which is oriented west–east, acts a bit like the much grander House of Dun (on pages 102–3). A visitor approaches the house in shadow but, on entering and passing through to the conservatory (c), reaches a zone of warmth and bright sunshine, with a view into a garden backed by hawthorns dark against the sun.

In the afternoon the situation is reversed (opposite, top). The entrance side of the house is sunny, the conservatory is in shade, and the garden is lit up by afternoon sun.

It is sometimes suggested that the best orientation for a conservatory is to face south. This is the direction in which sun and warmth is optimised. But in some ways an eastern orientation can be better (more subtle). Facing the morning sun, the conservatory is warmed early. But continued and excessive warming throughout the day is reduced by the shadow the house itself progressively provides. Also, in the afternoon and evening, when the conservatory is most used, the garden is lit up like a stage set.

In the morning, the front door of our house is in shadow while the garden is in sun. The sun also warms the conservatory.

LOOKING FROM SHADOW
evening shadow

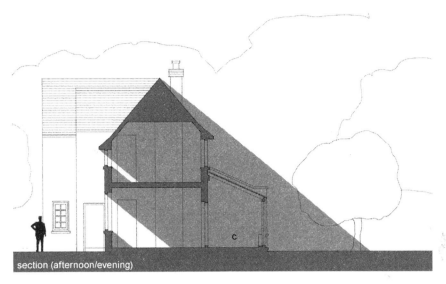

section (afternoon/evening)

While in the afternoon, the entrance is sunny and excessive warming of the conservatory is prevented by the shadow the house itself provides. And the view is into a sunlit garden rather than into the direct glare of the evening sun.

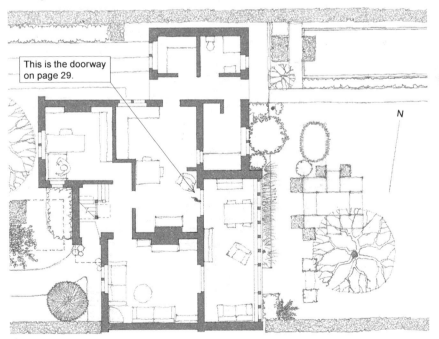

This is the doorway on page 29.

N

plan

DARK SIDE
House of Dun, William Adam

Shadow can affect your approach to a building.

House of Dun, aerial view

Building a grand country house in a temperate climate, you would want your principal rooms to benefit from the sun. In the northern hemisphere you would orient them to the south. Not wanting to reduce the amount of favoured sunny elevation available for living rooms, you would probably place the entrance on one of the shadier sides. Such an arrangement affects how your house is perceived by visitors. It has the additional possibility of composing an architectural sequence of spaces that leads from shade to sunlight: a sequence that affects your visitor's appreciation of the world in which you live.

This arrangement happens in William Adam's design for the House of Dun (above). A visitor is led around the house to be faced by the forbidding dark side, against the sun.

The House of Dun, near Montrose in Scotland was designed by William Adam in the 1730s and 40s. The approach is from the main road to the south but the driveway runs to the left of the house in the aerial view above and curves around (from a) to the main entrance in the northern elevation (b).

HOUSE OF SUN
shadow and sun

Shadow chills –
psychologically as
well as physically –
sun cheers.

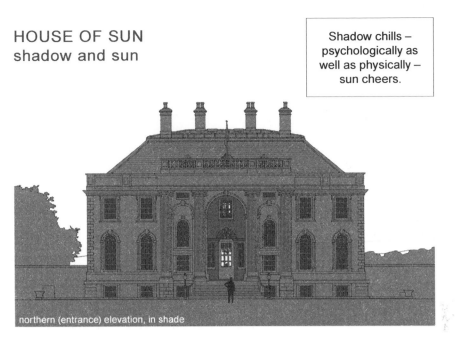

northern (entrance) elevation, in shade

Except in the Scottish summer's early mornings and late evenings, the entrance elevation of the House of Dun is in shade (above). It presents a forbidding face to the visiting stranger. The main doorway is in a double-height recess which not only exaggerates the status of the owner but shrouds the entrance in even deeper shadow.

The shadiness of the entrance elevation, silhouetted against the sun, is in contrast with the bright aspect of the southern (below). The windows of the principal rooms enjoy the sunshine as well as fine views towards the waters of the Montrose Basin, a few miles away.

The entrance to the House of Dun is almost always in shade…

… while the main rooms and gardens on the south are sunny.

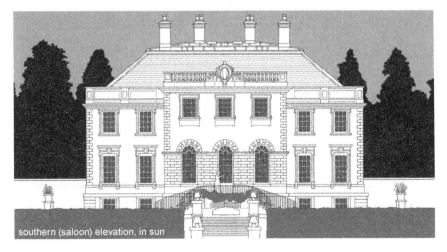

southern (saloon) elevation, in sun

LIVING IN A FAVOURED WORLD
an architectural sequence of spaces

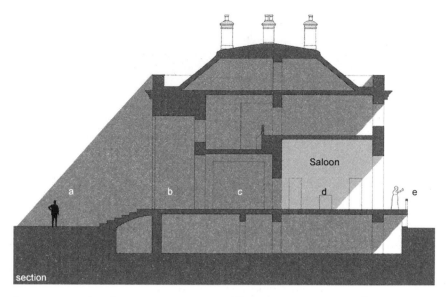

section

As you approach, enter and progress through the House of Dun, you move from shadow to sunshine. The contrast enhances the impression that the owner lives in a specially favoured world.

The sequence progresses from: the approach (a); up steps to the double-height porch (b); through the door into the hallway (c); through to the sunny Saloon (d); and out to look over the garden (e).

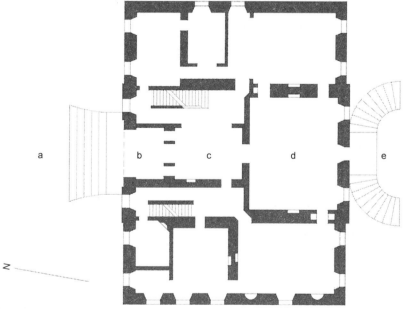

plan

FROM SHADOW INTO LIGHT
dramatic transition

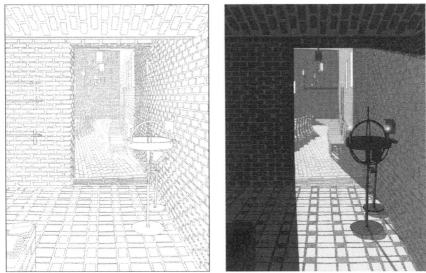

Church of St Petri, Klippan, Sigurd Lewerentz, 1963–6

Dramatic transition from light into darkness and vice versa is an important and powerful ingredient in the narrative potential of architecture. A world of monotonous illumination can be boring. Shadows intrigue, even though they might provoke trepidation. Light draws you out from shadow.

The glare from the windows in Lewerentz's church draws you from the shady chapel (above; see also page 88).

The light from the roof glazing over the stairs of Larsen's Glyptotek extension (see page 122) draws you out from its link from the original museum (below).

Glyptotek extension, Copenhagen, Henning Larsen, 1996

MORE DEATH
a crematorium

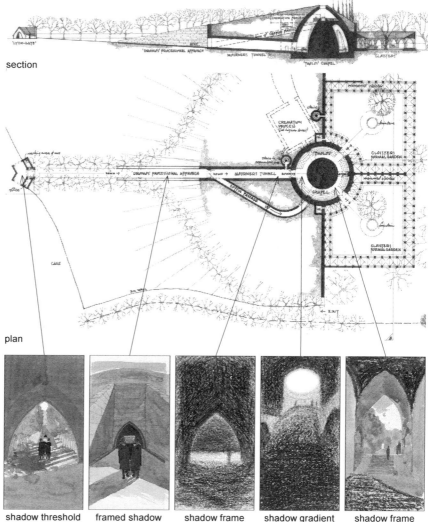

section

plan

| shadow threshold | framed shadow | shadow frame | shadow gradient | shadow frame |

Looking back at a project I designed as a student many years ago I realise that it has an underlying shadow narrative. This is appropriate since it was a design for a crematorium. The project seems to embody various metaphorical, historical and prehistorical allusions too. I am pleased it concentrated on using shadow as an instrument for orchestrating experience and emotion: a descent into gloom and an ascent into light.

My project embodies the processes of cremation and grieving. The funeral enters a deeply shaded chapel, like an ancient Mycenaean tholos, the interior of which is dominated by a huge and oppressive stone mound on the top of which, under an opening to the sky, the cremation takes place. After the ceremony the mourners proceed to cloisters, like Paradise, back outside in the sunlight.

REGIONAL SHADOW

The extremities of the earth, the north and south poles, spend large parts of each year in shadow. But between those two poles different latitudes experience different characters of light and shade, varying as the earth wobbles on its axis through the year, changing the angles at which the sun strikes its surface. Different regions have different climates, some hot and sunny with strong shadows, some disposed to cloudy overcast skies under which shadows are soft, almost non-existent. The earth spins as it wobbles, making night and day, but also those daily transitions between the two, dawn and dusk. The natural conditions for shadows – seasonal, diurnal, climatic, meteorological... – create infinite diversity. But architecture can exploit and modify them, celebrating or inverting regional predisposition.

ITALIAN SHADOW
Ivy de Wolfe's photographs

In 1963 the Architectural Press (London) published a book, *The Italian Townscape,* attributed to an author called Ivor de Wolfe (a nom de plume of Hubert de Cronin Hastings*). The book is replete with black and white photographs by Ivy de Wolfe (Hastings' wife Hazel). These illustrate the various themes identified as constituting the distinctive character of Italian towns. Although 'shadow' is not singled out as one of the themes identified, many of the photographs feature the deep shadows created by the Italian sun, as in the example I have redrawn above (from page 102).

'The borders between light and shade are distinct, dissect the piazza into irregular light and dark patches... I immerse myself in deep shade, only a few cornices are touched by the sun and cast sharp, narrow shadows onto the walls.'

Sharp-edged and deep shadows contribute to the character of Italian streets, and to the graphic power of photographic compositions.

* Artifice Books (London) republished this book in 2013 with an introductory essay by Alan Powers and Erdem Erten on Hastings' contribution to British architectural journalism.

Gerhard Wolf – 'Shadow Walk in Rome', in Brandi and Geissmar-Brandi, eds. – *The Secret of the Shadow*, 2002.

NORDIC SHADOW
of soft overcast, grey skies

The light, and hence the quality of shadows, is very different in northern climates where cloud cover softens and disperses the light from the sun. Shadows, rather than being stark and dark, are gentle and muted. The quality of shadows changes during the day and through the seasons… with prompts for romantic reverie (see quotation below).

Under the cloudy skies of the north the quality of shadows is very different from under the Mediterranean sun.

'I love to be out of doors at day's end… traces of the past are everywhere, from horse ponds glinting like mercury among the stubble fields to tiny labourers' cottages… their small-paned windows yellow-lit, woodsmoke curling from brick chimneys hundreds of years old. In the half-light of dusk, the old lanes empty of traffic, it's possible to leave behind the present day with its frightening uncertainties and enter a world in which heavy horses worked the land (and) the seasons turned with comforting regularity.'

Melissa Harrison – 'Nature Notebook', *The Times*, 16 November 2019.

WELSH SHADOW
cloud and low sun

Yng Nghymru mae cysgodion yn feddal.

Even when the sun shines in Wales the shadows are softer and with more subtle variations than those found under the bright sunshine of Mediterranean regions.

In northern climates shadows are softer – more modulated by cloud, moisture in the air, and low angles of sun – than in more southern countries where the sun is high in the sky and the air is clearer. Architecture is perceived differently in the light and shade of Wales, even when the sun is shining.

The sun and the deep shadow of Italy is more sympathetic to the 'patination' that Italians sometimes allow their buildings to acquire. The softer Welsh light and shade makes the wear and tear of buildings – cracked render and peeling paint – seem shabby and anything but romantic.

Architecture responds subliminally to the light in which it stands. Bright colours shine under clear skies with strong and stable sunshine. But under shifting cloudy skies muted colours harmonise better with the varying tonal gradations of softer shadows.

ANALYSING ARCHITECTURE NOTEBOOKS

TRANSPOSED SHADOW
Nordic Pavilion, Venice, Sverre Fehn

As architects we can play with the character of light in our buildings. Just as the architects of the television scenes of finished makeovers (see the following page) want to enhance the attraction of their improvements by suggesting the rooms have become sunnier, so too can more poetic architects seek to transport a particular climatic light to a different region.

A case in point is the Nordic Pavilion at the Biennale site in Venice, designed by Sverre Fehn in 1959. Here, in Italian light, Fehn replicates the evenly modified light of cloudy Scandinavian skies by means of a roof of deep concrete baffles running in both directions. These baffles, combined with fabric draped over them, prevent direct sunlight causing any sharp-edged shadows within the pavilion.

In his Nordic Pavilion, Sverre Fehn manages to transport the soft shadows of overcast Norwegian skies to the bright world of northern Italy.

SUBTLE DECEPTION
flattering a room with artificial (temporary) shadows

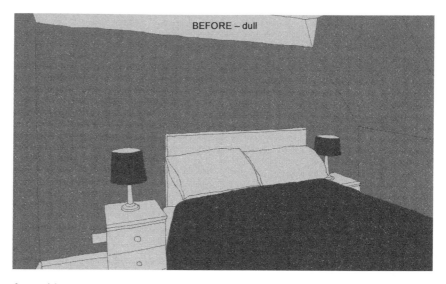

BEFORE – dull

As architects we are stage designers for actual lives. In stage design, light and shadow contribute to mood.

When writing this page (in June, 2019), the Hertfordshire village of Chorleywood was described as 'partly cloudy' on my weather app. A property there was the subject of an episode of the BBC TV series *Changing Rooms* in 2005.

The aim in the episode was to make the room more 'Mediterranean'... It was achieved mainly by the shadows cast by louvre screens positioned across the windows (below).

But the effect was enhanced by means evidently not always available in Chorleywood – sunlight – here provided for TV by a floodlight positioned outside the window.

AFTER – bright with shadows cast by 'Mediterranean sunlight' (provided by a TV floodlight)

STAGE SET FOR SHADOW

In a traditional proscenium arch theatre we, the audience, sit in shadow. While the action of the play or opera is brightly lit, the rest of our surroundings is turned to nothingness by absence of light. This is a powerful example of what I have, previously in this Notebook, called a shadow frame. By creating contrast it draws attention to the focus of the dramatic proceedings and accentuates the play of light and shade on the stage.

To some extent, architecture is always involved in the making of stage sets, sets on which the action of life takes place. Sometimes shadow frames are used to engineer pictorial contrast. Sometimes rooms or galleries are treated themselves as stage sets for the play of shadow. Shadow itself can become the subject of the drama.

SHADOW TYPES IN PAINTINGS
learning from artists

When the light and shade of a room catch the eye of an artist, architects can learn about relationships between shadow and atmosphere, and maybe try to emulate what they discover, in their own work.

The Dutch painter Rembrandt is well known for his chiaroscuro – the handling of shadow and light. This painting imitates Rembrandt's style. It is by a follower of his, Salomon Koninck, and dates from around 1645. The painting is a contained shadow lit from a single window. There is deep shade, shadow gradient and, along the corridor, a shadow threshold.

This small painting (c. 1859) by the Victorian artist Frederic, Lord Leighton, is a study of architectural types of shadow. It includes framed shadow and a shadow frame as well as cast shadow, a shadow threshold and deep shadow. (The painting is called 'Entrance to a House, Capri', and is displayed at Leighton House in London.)

Vilhelm Hammershøi, a Danish artist who died in 1916, made enigmatic paintings of

the interior of his house. Their underlying subject is light and shade.

SHADOW MYSTERIES AND LIGHT
subtlety and complexity

'The night gains an incredible value in that it erases space. The fire invents light and heat, and through this light darkness gains new importance: it creates the story, it entertains. The fire invents a room where there is light. The fire is a producer of space, and in the shadow, mystery is born.'

Alvar Aalto, quoted in Göran Schildt – *Alvar Aalto in His Own Words*, 1998.

In his notebook, Giacomo Leopardi muses on the subtleties of light obscured or modulated:

'It is shown how objects half seen, or seen with certain impediments, etc., awaken indefinite ideas in us and may explain why the light of the sun or the moon is pleasurable in a place where they cannot be seen and where the source of light is not revealed; a place only partly illuminated by that light; the reflection of this same light, and the various material effects that derive from it; the penetrating of this same light into places where it becomes uncertain and blocked, and is not easily made out, as through a reed bed, in a wood, through half-closed shutters etc., etc.; this same light seen in a place, object etc., where it does not enter and does not strike directly but is refracted on to them and diffused by some other place or object, etc., that it happens to hit; in a corridor seen from inside or from outside, and in a loggia likewise, etc., those places in which the light blends in, etc., etc., with the shadows, as beneath a portico, in a raised suspended loggia, between rocks and ravines, in a valley, on hills seen from the shade when their peaks are gilded; the reflection that, e.g., a coloured glass casts on those objects on to which are reflected the rays that pass through this same glass; all those objects, in short, that by means of different materials and slight changes in circumstance reach our sight, hearing etc., in a manner that is uncertain, hard to distinguish, imperfect, incomplete, out of the ordinary, etc. ... and for the reason identified above, the sight of a sky with a patchwork of little clouds, in which the sunlight or moonlight produces varied, and indistinct, and unusual, etc., effects, is also pleasurable. The same light is most pleasurable and very sentimental when it is seen in towns, dappled by shadows, where the dark contrasts in many places with the light, where the light in many parts fades gradually, as on roofs, where some secluded places hide the shining star from view, etc., etc. Variety, uncertainty, not seeing everything, and therefore being able to wander in one's imagination through things unseen, all contribute to this pleasure. The same goes for similar effects produced by lines of trees, hills, trellising, cottages, haystacks, uneven terrain, etc., in the countryside.'

A simple apse in a chapel can be an example of a shadow frame with metaphorical meaning – the offer of progress from mysterious shadow into the clarity of enlightenment.

Leopardi, trans. various – *Zibaldone: the Notebooks of Giacomo Leopardi* (early 19th century), 2012.

SHADING PLANES
Secular Retreat, Peter Zumthor

Shadows can be like musical chords.

An architect can
orchestrate shadow
like a composer
music.

early model of Chivelstone House (Secular Retreat), Peter Zumthor, 2010 (2018)

In 2010, Peter Zumthor was commissioned to design a house in South Devon. In its early stages it was called Chivelstone House but when complete, Secular Retreat (see also page 47). Even though later he suggested he was inspired by Andrea Palladio, it is clear from early models (above) that he also sought to emulate the modulated shade of prehistoric dolmen structures.

This model seems to have been constructed as a piece of three-dimensional architectural music, with careful consideration being given to the ways surfaces and openings orchestrate tone and shade.

'The shadow gives shape and life to the object in light. It also provides the realm from which fantasies and dreams arise. The art of chiaroscuro is a skill of the master architect too. In great architectural spaces, there is a constant, deep breathing of shadow and light; shadow inhales and illumination exhales light.'

Juhani Pallasmaa – *The Eyes of the Skin: Architecture and the Senses* (1996), 2005.

SHADOW ORCHESTRATION
installation, Grafton Architects

Grafton Architects' 'Sensing Spaces' installation consisted of a composition of flat panels defining a lofty functionless space. Orthogonal but irregular, it was lit mysteriously from above.

The light from the mysterious source reflected off the flat surfaces in subtle ways that changed during the day and in different weathers. The installation was a wordless architectural poem on the subject of shadow.

An exhibition of architectural installations was held in the galleries of the Royal Academy in London in 2014. It was titled 'Sensing Spaces'. The Irish practice Grafton Architects – Yvonne Farrell and Shelley McNamara – contributed a piece on the subject of shadow.

They created a lofty space sculpted by a complex but generally orthogonal timber structure of flat off-white panels suspended from the high gallery ceiling. The purpose of this structure was to act as an instrument for the orchestration of light and shade. The result was the architectural equivalent of a very slow piece of ambient music, the identification of a place in which shadows rotated and morphed through the day as the sun moved and the weather changed. The piece reinforced the observation that architecture is not always led by function, nor by sculptural form. Its subject can be light and shadow changing with time and external conditions.

IF WE HAD NO SHADOW...
Barcelona Pavilion, Mies

Shadow may be needed to complete a building's narrative.

Mies van der Rohe's Barcelona Pavilion (1929; see also page 72) is primarily a composition of independent planes orthogonally arranged in the three dimensions (right). It has also been discussed as an essay in reflection.* It is a work of light and shadow too.

The above drawing shows part of the pavilion depicted in a shadowless world. It focuses on the statue 'Dawn' by George Kolbe, a contemporary of Mies.

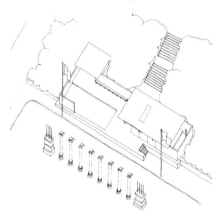

* Robin Evans – 'Mies van der Rohe's Paradoxical Symmetries', in *Translations from Drawing to Building and Other Essays*, 1997.

See also *Twenty-Five Buildings Every Architect Should Understand*, 2015, pp. 25–42.

COMPLETING THE NARRATIVE
'Dawn', George Kolbe

Light is an ambient condition of architecture... as is time. The shadows cast by the sun as it moves across the sky combine both.

In Mies van der Rohe's Barcelona Pavilion the narrative of the sculpture is only completed when light and shade are admitted. The sculpted lady, who might have been dancing, is revealed as shading her eyes from the light. Her form is framed by the shadowed rectangle created by the pavilion's roof and walls. By the morning sun she acquires her shadow companion projected onto the wall behind her. But during the day and into the evening she returns to the shadows... and prepares to meet her namesake again the following morning.

Shadow, as exploited here by Mies, ties architecture into grander rhythms...

The narrative of Kolbe's 'Dawn' is only completed by Mies's architecture which orients the morning sun to shine into her face – which she shades with her hands – and provides a screen wall onto which her shadow is projected.

STAGE SET
Canova Museum, Carlo Scarpa

Shadow play
enlivens a stage
for exhibition.

This is the view from one of the original galleries of
the Canova Museum in Possagno, Italy (designed
by Francesco Lazzari in 1836 to house the work
of sculptor Antonio Canova) into an extension
designed by Carlo Scarpa in 1957.

*The section (below) shows
the original galleries to
the right and Scarpa's
extension to the left.*

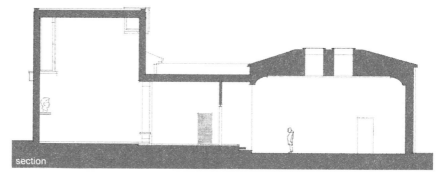

section

PLAY OF SHADOW
illuminating and seasoning

The extension was attached to the west side of the original gallery. The contents, being plaster casts and sculptures, are (like those in Sir John Soane's Museum; see page 85) generally monochromatic. They are not light sensitive but, being three-dimensional, need light to be displayed at their best. The lie of the land suggested the floor level of the new addition be slightly higher than that of the original gallery. The topography helped Scarpa design one of his galleries as a stage (see the section, opposite) upon which the sculptures could be displayed (above). This gallery is lit by idiosyncratic inset windows high in its upper corners. These produce, as befits a stage, a shadow play that changes through the day, seasoning and adding variety to the way the exhibits are displayed.

The arrangement of windows, projecting into the space at its corners, creates an ever-changing pattern of shadows slanting across the walls and sometimes illuminating the exhibits. In this way, the gallery has more vitality than if the exhibits were lit only by constant electric lights.

CHANGING PATTERNS
Glyptotek, Henning Larsen

Shadow play as art installation.

In 1996, Henning Larsen completed an extension to the Glyptotek Museum in Copenhagen (see also page 105). The building consists of a shallow stair wrapped around a square tower of galleries. The tall and narrow staircase space is lit from above. On sunny days the regular structure of the glazed roof projects a changing pattern of sunlight and shadow onto the walls and across the stair (below left).

The staircase space of Larsen's Glyptotek extension (left) could have been merely practical – a means of getting from one floor to the next. Because of the play of sunlight, changing slowly but constantly during the day, the space becomes itself a grand architectural work of art, an installation whose subjects are sunlight, time and shadow.

FROM SHADOW TO LIGHT
Glasgow School of Art, Mackintosh

A classic architectural strategy is to lead people from shadow to light.

Glasgow School of Art was designed in the 1890s. The entrance elevation (below), with its large windows, faces north, so that the studios receive evenly shadowed light. But early on clear summer mornings the modelling of Charles Rennie Mackintosh's entrance is illuminated in sunlight.

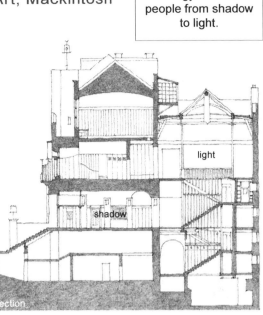

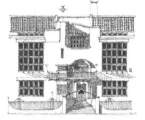

Mostly, however, one enters through a shadowed elevation, passing into a low-vaulted lobby (see the partial plans below and the sketch below right). From there you can see, through a rectangular archway, a stair leading you upwards.

The stair offers the prospect of passing from the shadows of the entrance lobby to the sky-lit museum/exhibition space on the first floor. In this way you follow a classic example of a route from shadow up into light.

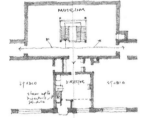

upper floor plan

entrance floor plan

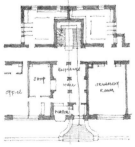

SHADOW IN FILM
contributing to narrative and drama

Cinema can teach architects about the dramatic potential of shadow.

Films depend on architecture for their settings. Being photographic, their cinematographers and directors often seem to pay more attention to shadows than architects designing buildings for the real world. In some styles of film, making shadows is particularly important, often being used in narrative and symbolically charged ways. Two examples, from the era of black and white and film noir are illustrated on this page. They are the 1943 version of Charlotte Brontë's *Jane Eyre*, and the 1949 film adaptation of Graham Greene's novel *The Third Man*. Both starred Orson Welles, who had also used shadows to powerful narrative and dramatic effect in his own film *Citizen Kane* (1941).

The contribution shadows make to the narrative of films is pictorial, metaphorical and symbolic. Pictorial, metaphorical and symbolic shadows can contribute to the non-literary narratives of 'real-world' architecture too.

* See also pages 49–54 of the *Metaphor* Notebook.

Jane Eyre, Robert Stevenson, 1943

In the film of Charlotte Brontë's Jane Eyre *Rochester (Orson Welles) and Jane (Joan Fontaine) are about to be married. The vicar asks if anyone knows of any impediment. Above the altar, in the frame of an arch, a window's shadow forms a cross. As the vicar asks his question, the door of the church, off-screen, opens, its light casting the ominous shadow of a person on the wall opposite. He has come to declare an impediment…*

The Third Man, Carol Reed, 1949

At the beginning of The Third Man *(written by Graham Greene), an American writer Holly Martins (Joseph Cotten) is in Vienna seeking his friend Harry Lime (Orson Welles). Martins hears that Lime has, allegedly, been killed in a road accident and attends his funeral. Later, as he is leaving Vienna, Martins (helped by a kitten) becomes aware of a figure concealed by the deep shadow of a doorway* in a war-scarred building. A flash of light from a neighbouring window reveals it to be Lime.*

ANALYSING ARCHITECTURE NOTEBOOKS

SHADOW AND TIME

The sun moves across the sky, from east to west. And as it does, shadows, moving in counterpoint, change in size and character. For thousands of years we have told the time by the movement of the sun's shadows. We have used architecture to build instruments for doing so. The atmosphere of places is changed by the diurnal rotation of shadows. The slow movement of shadows through time is part of architecture. Its stimulating variability is destroyed by the constancy of electric lighting.

'The shadows fall across the long-nosed sundial and tell me that I exist, I exist...'

Lawrence Durrell – *Spirit of Place*, 1969.

'Cut my shadow from me.
Free me from the torment
of seeing myself without fruit.
Why was I born among mirrors?
The day walks in circles around me,
and the night copies me
in all its stars.'

Federico García Lorca, trans. Merwin – 'The Song of the Barren Orange Tree' (early 1920s), 1955.

SUNDIAL ARCHITECTURE
hemicyclium, Carthage

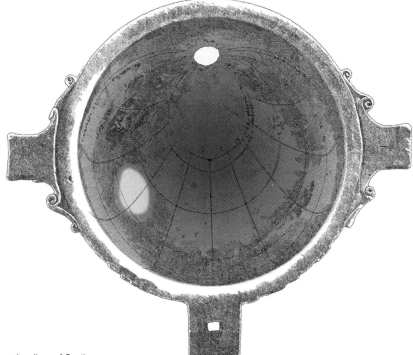

hemicyclium of Carthage

The hemicyclium of Carthage (c. 1st century CE; in the Louvre, Paris) is a roofed spherical sundial. It consists of a hemispherical bowl tipped forward directly towards the south. At its top is a hole through which the sun sends a beam of light penetrating the bowl's inner shadow. Through the day (and the seasons) the angle of the sun changes. The position of the sun's light patch on a matrix incised into the bowl's inner surface indicates the time.

This instrument is already 'architectural' in that it identifies a place for telling time. It also has a 'roof' that creates a contained shadow which the sun's beam penetrates. If we were, in our imaginations, to complete the inner sphere by adding another matching hemisphere (opposite above), and make it large enough to go inside, we can imagine the experience of witnessing a sunbeam sweeping slowly around in the darkness like a searchlight, marking time.

It is likely that such sundials were inspired by seeing the movement of shafts of sunlight admitted into shady rooms by small windows or roof openings. But such sundials themselves inspire grander works of architecture.

The Pantheon in Rome (2nd century CE), with its spherical geometry and oculus at the highest point of the dome, is effectively a roofed spherical sundial. The movement of the sunlight entering through the roof opening establishes a link between the internal space and the moving sun.

ANALYSING ARCHITECTURE NOTEBOOKS

'The cold domed room of the tower waits. Through the barbicans the shafts of light are moving ever, slowly ever as my feet are sinking, creeping duskward over the dial floor. Blue dusk, nightfall, deep blue night.'

James Joyce – *Ulysses*
(1922), 1968.

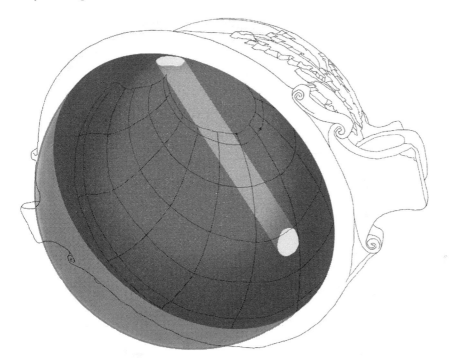

the interior sphere completed

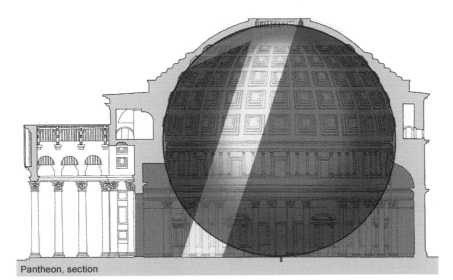

Pantheon, section

SEARCHLIGHT
Il Gesù, Rome, Vignola

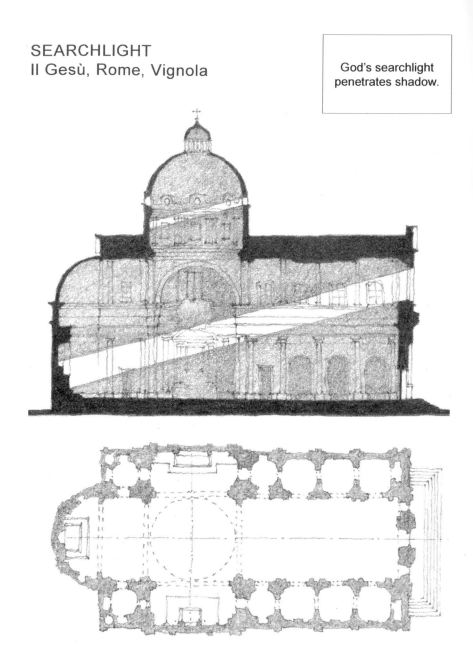

If the sun is shining and you are there at the right time of day (mid to late afternoon), the drama of the interior of the Baroque church of Il Gesù in Rome (Vignola, 1568–75; not very far from the Pantheon) is vastly increased by a shaft of bright sunlight striking through the darkness. Originating in the celestial sun, it enters the church through a clear glass window high in its western gable. Slowly the beam tracks across the sanctuary. For a moment it may illuminate the altar directly. Is it the searchlight of a watching God penetrating the shadows?

A shaft of sunlight enters the shady interior of Il Gesù in Rome. It establishes a direct link with earth's ultimate source of power.

LINKING ARCHITECTURE WITH THE CELESTIAL
Can Lis, Mallorca, Jørn Utzon

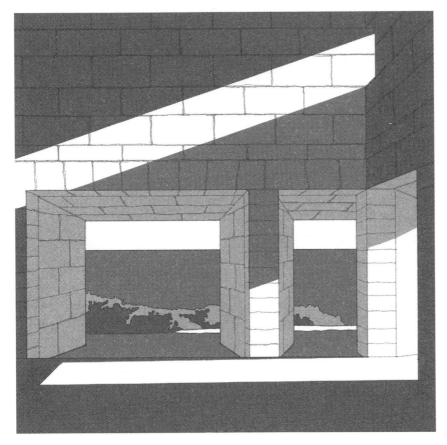

There is a celebrated example of time-related shadow in the house Jørn Utzon built for himself on the Mediterranean island of Mallorca. Richard Weston describes his experience of it below.

'Sunlight fills the openings but all around the walls are gathered in shade. But then, in the mid-afternoon (it was at 3.40pm when I was last there in March 2000), a slice of sun, so distinct you feel you could pick it up, arrives on the floor through a small glazed opening – too rudimentary to call a "window" – placed high in the west wall. After a minute or two the sun catches the slightly raised mortar and cut edges on the stone... Moments later and a diagonal slash of wall above the windows is dusted with light. This intensifies for a few precious minutes into a stone-dissolving shaft and twenty minutes later begins to recede, leaving only a glowing patch of light to linger in the opening almost until sunset as a reminder of the sun's daily visit.'

A shaft of sunlight enters the shady interior of Can Lis, Jørn Utzon's house on the island of Mallorca. As a sundial it only 'tells the time' during one short period each day... but it marks that moment with affecting shadow play.

Richard Weston – *Utzon*, 2002.

TELLING THE TIME
Piazza S. Pietro, Rome

Piazza S. Pietro
is a sundial.

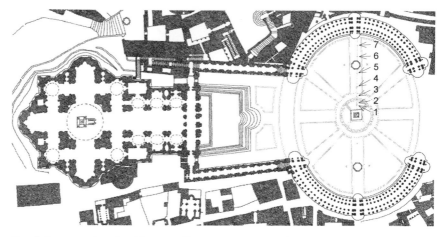

Sundials are works of architecture. They are instruments for telling the time. They also identify places for making sense of the progress of time, through the day and through the seasons of the year.

More sophisticated works of architecture can be time pieces too. The sun moves around them and the shadows they cast move and change length during the day and the seasons of the year.

With is central obelisk, the Piazza S. Pietro in Rome is a clock, with shadows for its hands.

As well as framing a place of assembly, where crowds of people can be embraced by Bernini's curving colonnades, the Piazza in front of St Peter's in Rome is also a sundial. At its centre stands an 84 foot tall obelisk brought originally from Egypt. The obelisk acts as a gnomon. As the sun moves during the day its shadow changes in length and angle. There are seven metal plaques along the north–south meridian (above). These indicate the length of the obelisk's shadow at noon on specific days of the year.*

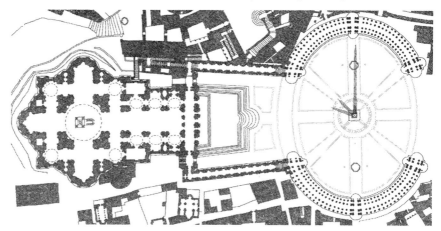

The shadow reaches plaque 1 on 21 June, the longest day, and plaque 7 on 21 December, the shortest.

* See page 34 of the *Metaphor* Notebook.

RHYTHM OF THE DAYS
Palace of Charles V, Granada

You do not need a gnomon for the movement of shadows to show the passage of time.

On the same hill as the Islamic palace of the Alhambra (in Granada, Spain) stands the neoclassical palace of Charles V (1527, but not roofed until the twentieth century). At the core of its square plan (right) is a circular courtyard lined with two storeys of shaded columned walkways.

During daylight, this circular courtyard is a clock with no hands. As time passes the sun illuminates a progression of sectors of the courtyard and the shaded walkways.

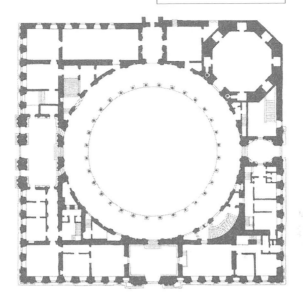

The Charles V Palace in Granada (by Pedro Machuca), which was never occupied as a palace, is, when the sun shines, an architectural clock. A time lapse film from vertically above its courtyard would show shadows revolving clockwise around the circular space. They trace paths that vary with the seasons of the year too.

All sunlit architectural space has this temporal dimension on which shadows tell time.

SHADOW PLAY
Gálvez House, Luis Barragán

A private domestic shrine to moving shadow.

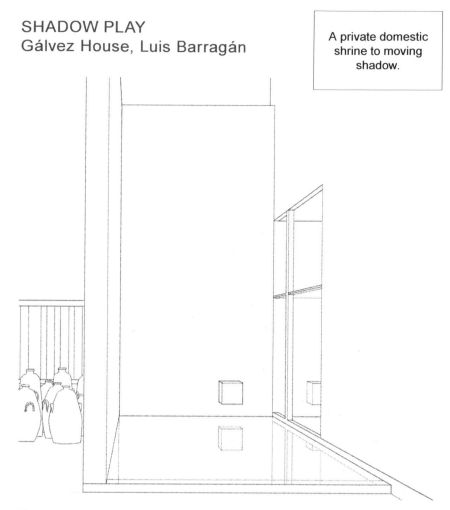

The small courtyard of the Gálvez House without shadow...

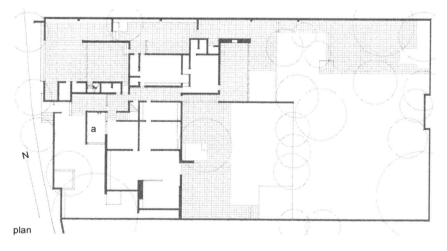

plan

SHADOW PLAY
Gálvez House, Luis Barragán

… and with… (in the afternoon). (Mexico City is at latitude 19.4° N.)

'The living room reaches out through an enormous window into a small courtyard. Two pink walls mark its boundaries, acting as the fourth wall of the living room and transforming the courtyard into an object of abstract contemplation. The space fills the room with its gentle atmosphere and expresses the essence of the entire house without any marked functionality – it contains only water, walls and light.'

Raúl Rispa, ed. – *Barragán: the Complete Works*, 1996. (This book also contains the photographs on which the drawings on the following page are based.)

The courtyard referred to in the quotation alongside is a small part of a villa in Mexico City, the Antonio Gálvez House, designed by Luis Barragán in 1955. The courtyard is indicated at 'a' in the plan opposite. The courtyard is open towards the south, so it catches sunlight (sometimes dappled by the leaves of adjacent trees) during most of the day.

If I were allowed, I would change one word in the quotation's last sentence: from 'light' to 'shadow'. As can be seen from the sequential drawings on the next page, this small pink courtyard, with its reflective glass and pool, is a scene of constantly changing shadow play. Capturing this is what makes this small episode of architecture into 'an object of abstract contemplation'.

KINETIC SHADOW
Gálvez House, Luis Barragán

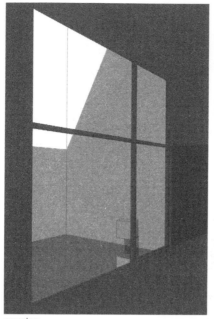

morning

late morning (around noon)

early afternoon

afternoon

(the image on the previous
page is later in the afternoon)

JAPANESE SHADOW

The aesthetic contribution of shadows to traditional Japanese architecture is described in Junichiro Tanizaki's 1933 work translated as *In Praise of Shadows*, quotations from which are included in the following pages. I have tried here to describe some of the many ways in which shadows contribute to the appearance and atmosphere of Japanese spaces – rooms and gardens – and to do so in a manner that might prompt ideas for experiment in your own design work.

Tanizaki's book is a hymn to what he sees as a threatened world of (Japanese) shadow – threatened by the seduction of bright (Western) lights. The final words of the book are quoted below.

*'I would call back... this world of shadows
we are losing... I would have the eaves deep
and the walls dark, I would push back into the
shadows the things that come forward too
clearly... Perhaps we may be allowed at least
one mansion where we can turn off the electric
lights and see what it is like without them.'*

Junichiro Tanizaki, trans.
Harper and Seidensticker
(1954) – *In Praise of Shadows*
(1933), 2001.

PRAGMATIC SHADOW
waiting shelter

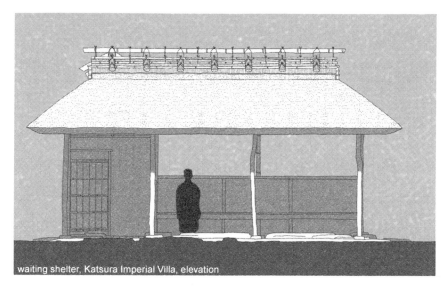

waiting shelter, Katsura Imperial Villa, elevation

The waiting bench provides a shady refuge from which to admire the prospect of the gardens.

Traditional Japanese architecture uses shadow in practical and aesthetic ways. Guests to a tea ceremony are invited to sit on a bench that shelters them from rain and sun while they wait for the host to guide them to the tea house (see pages 140–45). From within the shadow container created by its roof they may contemplate the gardens around.

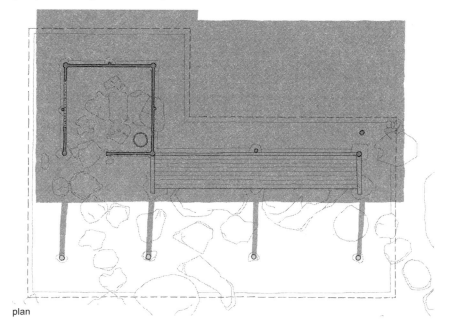

plan

ANALYSING ARCHITECTURE NOTEBOOKS

SHADOW PROJECTION
moving patterns on a wall

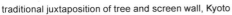

traditional juxtaposition of tree and screen wall, Kyoto

The gentle play of the shadows of a tree on an adjacent screen wall (above) supplements the visual interest of a Japanese garden. The barrier wall, which could appear harsh and implacable acquires charm by the patterns and movement of the shadows cast by the leaves and branches of a purposely placed tree.

Shadow may not be produced by forces under our control; we cannot extinguish the sun nor prevent its passage across the sky. Nevertheless, we can exploit its potential for adding visual interest to the places we make. Traditional Japanese architects employed a variety of strategies for doing this in their gardens as well as their buildings. These strategies can be employed in contemporary architecture too.

The Japanese architect Tadao Ando used this device in his design for the Conference Pavilion at the Vitra Campus at Weil am Rhein in Germany (left). In this case the wall is also given a faint geometric pattern by the traces of the formwork in which the concrete was cast. This creates a classic combination of fixed (human intellectual) geometry with the irregularity and movement of (natural) vegetation and shadow.

Conference Pavilion, Vitra, Tadao Ando, 1993

SHADOW PROJECTION
the pattern of sunlight through a screen

Shoka-tei

In traditional Japanese architecture screens are sometimes constructed and oriented primarily for the shadows they will project onto interior surfaces.

The example above is in the Shoka-tei, one of the tea houses in the meticulously kept stroll gardens of the Katsura Imperial Villa on the outskirts of Kyoto. Further examples from these gardens are illustrated on the following pages.

Engineered shadows add visual interest to a space, especially spaces that are dedicated to pleasure and meditation rather than work.

SHADOW THRESHOLD
Imperial Gate, Katsura

section

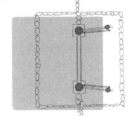

plan

The Imperial Gate at Katsura defines its threshold in multiple ways. The gate posts are planted in a slightly raised platform defined by a perimeter of stones. Between the posts is a raised threshold over which the visitor must step. The entrance transition is extended by the open gates and the stops against which they rest. The gate posts support a thatched roof which, together with the posts, platform and threshold, frame the moment of entrance.

The roof casts a shadow through which visitors must pass. This is perhaps the threshold definition that is least noticed but it is in some ways the most significant. Passing through the gateway you are washed by shadow and enter the Imperial estate cleansed.

SHADOW CONTAINER
Shokin-tei, Katsura Imperial Villa, Kyoto

In the gardens of Katsura stand various pavilions. One is the Shokin-tei, a tea house built in the seventeenth century. It stands by a lake and faces west and north-west. Its principal spaces enjoy the afternoon and summer evening sun filtered through trees. The roof is thatched with deep eaves so that the interior rooms are almost always in shadow. Looking at the pavilion across the lake (above) the interior is in deep shade which envelops and conceals those inside, making the pavilion a carefully positioned refuge with a panoramic prospect of the carefully composed landscape around. Getting closer you see that the building is a sophisticated instrument for the subtle manipulation of that interior shade. (See following pages.)

'Even at midday cavernous darkness spreads over all beneath the roof's edge, making entryway, doors, walls, and pillars all but invisible.'

The Shokin-tei is another of the pavilions in the magnificent stroll gardens of Katsura Imperial Villa. Its section and plan (opposite) show it to be a sophisticated instrument for managing light, shade and views of the surrounding landscape. One can imagine delicate plucked-string music drifting across the lake from its shadows.

Junichiro Tanizaki, trans. Harper and Seidensticker (1954) – *In Praise of Shadows* (1933), 2001.

INSTRUMENT FOR MANIPULATING SHADOW
Shokin-tei, Katsura Imperial Villa, Kyoto

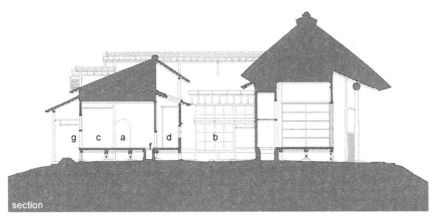

section

The roofs of the Shokin-tei establish a refuge of deep shade and shelter from which the beautiful gardens can be viewed. Views from the interior are managed – responding to seasons and weather – by translucent shōji screens.

A small but important part of the pavilion is the tea room (at 'a' in the drawings, and illustrated on the following pages).

In the centre of the building is a tiny courtyard (b) which allows some light into the heart of its plan.

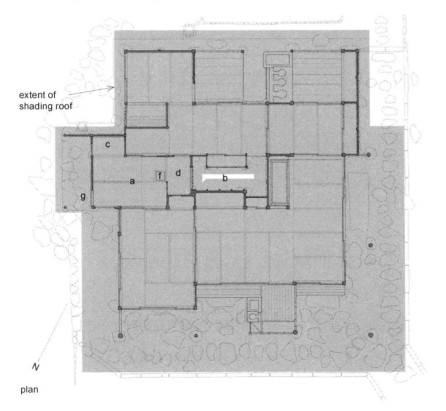

extent of shading roof

plan

SPATIAL MATRIX OF A TEA ROOM
even before light is admitted

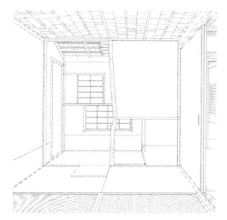

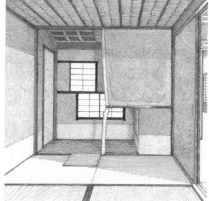

The tea room of the Shokin-tei is a carefully composed spatial arrangement. Its proportions are regulated by the standard size of tatami mats. The host lobby is separated from the main part by a screen supported by a post whose natural form contrasts with the room's rectangular geometry. Its architecture frames each part of the ceremony.

As well as being a carefully composed spatial arrangement, the tea room is a receptacle of shade counterpointed with light and, eventually, views into other rooms and the surrounding landscaped garden. The shady composure of the room is further enhanced by the dusky patination of its walls and the woven birchwood ceiling.

The tea room of the Shokin-tei is a three tatami mat space plus a tokonoma (at c in the drawings on the previous page) and an almost full width host entry lobby (at d; shown above) with a door leading from the private preparation spaces. A square hearth (f; covered in the drawing above) is set into the floor near the host lobby. Guests enter the tea room through a crawl doorway (at g in the drawings on the previous page) which means they must sacrifice some dignity in the interests of casting off pride and any sense of superiority over other guests.

After the ceremony, guests would be invited (through the sliding doors shown open to the right in the drawing above) into the other rooms of the pavilion with their fine panoramic views across the lake and the rest of the gardens.

The tea room of the a Shokin-tei is lit in various subtle ways. Its entrance faces the north-east and so is almost always in shade. Light inside is provided from openings above the crawl doorway, one of which may be closed with a translucent shōji screen. During the ceremony, while the room is closed, the whole space is in dusky shade counterpointed by the glow from the shōji screens in the host lobby (shown above) which receive some light from the pavilion's tiny central courtyard.

After the ceremony, when the sliding doors to the rest of the pavilion are opened, the shade is replaced by brighter light and the guests' focus is drawn, when further screens are opened, out of the capsule of the tea room to the exterior world of the host's meticulously composed gardens.

SHADOW CONTAINER...
Mongyo-tei, Kyoto, Japan

The Mongyo-tei (right) is another traditional Japanese tea house, set on the bank of a small lake in the stroll garden of Hakusasonso (the villa of an early twentieth-century artist, Kansetsu Hashimoto). It was used for meditation.

There are many subtleties to the composition of this small tea house* but here we are concerned with the various ways in which it exploits shadow. The first is that its roof creates a shadow container.

The most prominent element of this tea house is its pyramidal thatched roof with deeply overhanging eaves. This creates, on sunny days, a pool of shade within which the tea room sits. On wet days it is a dry refuge, but one from which you can contemplate the rain dripping from the eaves, rippling the surface of the lake, making rocks and greenery glisten...

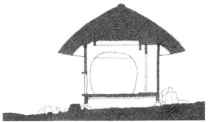

section

shadow container created by the roof

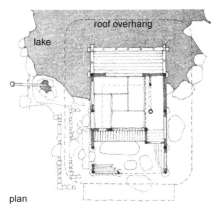

plan

The fundamental motivation of all architecture is to identify a place. The purpose of this tea house is to identify a place for calm reflection as well as for performance of the tea ceremony. This is achieved primarily by means of a roof that establishes a place of shade and shelter. The roof creates a shadow container within which sits the place for the tea ceremony or meditation. The primary adjectives that identify this place are 'shady' and 'dry', characteristics that provide conditions suited it to its purpose.

The tea house sits on the bank of a small irregular lake.

* See Case Study 12 in *Analysing Architecture*, 4th ed., 2014, pp. 298–308.

'In making for ourselves a place to live, we first spread a parasol to throw a shadow on the earth, and in the pale light of the shadow we put together a house.'

Junichiro Tanizaki, trans. Harper and Seidensticker (1954) – *In Praise of Shadows* (1933), 2001.

Obstructing the sun (and the rain) the large thatched roof establishes a shadow container. This would be a place of refuge in its own right, a place in which you can feel secure, protected from the elements and hidden in the shade. The tea room orchestrates this place of shadow.

The platform floor, the walls and the openings of the tea room have, conceptually, been slipped into the shadow container established by the roof. They organise and manage the place of shade and shelter in subtle ways creating an architectural poem.

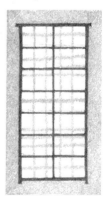

The tea house uses shadow in different ways too. Because of the relative shade inside the tea room, the view of another larger tea house across the lake is provided with a shadow frame – softened by its curved timber outline – which intensifies the sunlight and colour outside.

Also, the tea house is provided with a number of 'windows' 'glazed' with paper. The one shown above (from both sides) is within the tea room (visible at the left in the drawing on page 143). Thin matrices of timber or bamboo create counterpoint patterns of frame and shadow.

SHADOW FRAME AND FRAMED SHADOW
intensification and mystification

shadow frame

framed shadow (same window)

Shadow frames and framed shadows are often reciprocations of each other. The view into a garden from inside a dark room is intensified by the circular shadow frame created by a hole in the wall (above) whilst the same hole from outside (above right) draws attention to the mysteriousness of the dark interior.

Some Japanese rooms are so dark (below) that their boundaries – walls and corners, ceilings and floors disappear into gloom. The glare from the windows, which can be open or screened by shōji, serves only to intensify the shadow. The result is an immersive experience apt for a meditative ceremony.

the unapologetic gloom of a traditional tea room

SHADOW PROJECTION
wall 'paintings' that change during the day

shadows of a rough external screen on a more precisely framed paper shōji

In our contemporary architecture much effort is often made to engineer even and optimum light in all spaces (see the quotation on page 10). Shadows that reduce our vision can be annoying and are potentially dangerous. But traditional Japanese architecture shows that shadows can contribute to the aesthetics of architecture. The examples shown on this page are not shadows created by artificial lighting. They are achieved by the subtleties of architecture – the careful placing of shōji in relation to sources of light and other elements of the building, working in concert with internal contained shadow.

Paper shōji, held in the geometry of carefully crafted timber frames, provide effective screens for the projection of shadows (above and below).

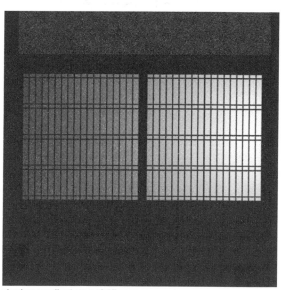

shadow gradient on a shōji screen

STAGE SET FOR SHADOW
tokonoma

'Our ancestors made of woman an object inseparable from darkness, like lacquerware decorated in gold or mother-of-pearl. They hid as much of her as they could in shadows.'

'In the still dimmer light of the candlestand, as I gazed at the trays and bowls standing in the shadows cast by that flickering point of flame, I discovered in the gloss of this lacquerware a depth and richness like that of a still, dark pond, a beauty I had not before seen.'

'So dark are these alcoves, even in bright daylight, that we can hardly discern the outlines of the work; all we can do is... tell ourselves how magnificent a painting it must be.'

'A Japanese room might be likened to an inkwash painting, the paper-paneled shōji being the expanse where the ink is thinnest, and the alcove where it is darkest. Whenever I see the alcove of a tastefully built Japanese room, I marvel at our comprehension of the secret of shadows, our sensitive use of shadow and light.'

Junichiro Tanizaki, trans. Harper and Seidensticker (1954) – *In Praise of Shadows* (1933), 2001.

PAINTING WITH SHADOW
inside a sequence of Japanese rooms

'And so it has come to be that the beauty of a
Japanese room depends on a variation of shadows,
heavy shadows against light shadows – it has
nothing else.'

Junichiro Tanizaki, trans.
Harper and Seidensticker
(1954) – *In Praise of Shadows*
(1933), 2001.

ANALYSING ARCHITECTURE NOTEBOOKS

SHADE REFUGE FOR CONTEMPLATION
Ryōan-ji temple, Kyoto

The architecture of the temple of Ryōan-ji on the outskirts of Kyoto sets up a relationship between the person and its famous rock garden. The veranda of the temple is a viewing platform where you can sit and contemplate the composition of boulders in raked gravel. The focus is on the rock garden, but the shady veranda is essential to the transaction between that focus and the person contemplating it.

'Our garden aims... at the miniature representation of Nature's grand scenery, and so the close-set little objects naturally produce dusky shades. However, this is not the only cause of shadiness in our garden. Dusk is a feature of Japanese artistic taste, as is seen in other arts of ours such as India-ink painting or the setting for the tea ceremony. Fundamentally analysed, it comes from Zen Buddhist philosophy, where inward peace and enlightenment are sought in Nature's twilight profundity. It is a tendency of the mind quite opposite from the love of the colourful, the bright, the obvious in Nature.'

Tsuyoshi Tamura – *Art of the Landscape Garden in Japan*, 1936.

rock garden, Ryōan-ji Temple, Kyoto, 1450

That relationship is enhanced and intensified by the shadow enveloping the veranda, creating a contrast with the sunlit rock garden and providing the person who contemplates it with the refuge of shade.

ONE MODERN EUROPEAN EQUIVALENT
Beyeler Gallery, Basel, Renzo Piano

Many European and American architects have sought to emulate the charm and aesthetic quality of traditional Japanese architecture in their own work. From Frank Lloyd Wright to Renzo Piano, there are far too many instances of this influence to cite here.

Just as one example, the illustrations on this page are of a gallery near Basel in Switzerland designed by Piano in the late 1990s. Not only does this building create a shadow container with its overhanging roof, it also transfers the soft light provided through the paper shōji screens to its sophisticated roof, which has various translucent layers that shade and soften the light from the sky as well as keeping out the weather.

Beyeler Gallery, Renzo Piano, 1997

The purpose of the Beyeler Gallery near Basel in Switzerland is to accommodate artworks in protective shade but with light enough to see them. The gallery stands in its landscape like a large contemporary tea house dedicated to art, a shadow container from which you can see the sunlight on the gardens around. And, as in the Shokin-tei (pages 140–42) the way in which it admits light aims to create (to use Tsuyoshi Tamura's phrase on page 149) a 'dusky shade' – an even light without the drama (and potentially damaging effects) of direct sunlight. To do this the Beyeler has a layered roof that acts like a horizontal translucent 'shōji' screen (see the section below).

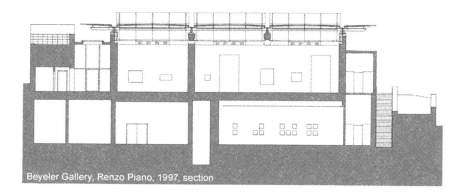

Beyeler Gallery, Renzo Piano, 1997, section

ANALYSING ARCHITECTURE NOTEBOOKS

ISLAMIC SHADOW

Traditional Islamic architecture originates in the hot and sunny countries of the Middle East. In such climates, shadows are an important element in making spaces comfortable for inhabitation. Quintessential examples of Islamic secular architecture include the souk, the walled garden, the han or caravanserai, the courtyard house and the hamam (bath house). All are places where shadow is an essential element. Shadows contribute to the aesthetics of Islamic architecture too. Islamic artists have created intricately carved low relief panels of scripture, verse and of complex tessellated geometric patterns, all of which enlist shadow to achieve their effect. Pierced screens, also with intricate geometric patterns, reduce the amount of sunlight entering rooms and decorate those rooms with cast shadows. Labyrinthine palaces and their harems are tapestries of shade...

'God... hath rewarded their constancy, with Paradise and silken robes: Reclining therein on bridal couches, nought shall they know of sun or piercing cold: Its shades shall be close over them, and low fruits shall hang down.'
Koran: 76, trans. Rodwell (1909), 1992.

DRAWING/WRITING WITH SHADOW
inscription and pattern

repeated 'There is no Victor but God' inscription, Alhambra, Granada, Spain

The wall surfaces of great Islamic buildings are often highly decorated with colourful geometric tile work and intricate low relief carving, the latter using shadow to draw. Avoiding depiction of human and animal form, these carvings, which involve much hand labour, have various themes. Three are illustrated on this page: the recitation of prayer inscriptions and poetry (above); complex interwoven geometric patterns (right); and stylised plant form patterns (below; in which no two segments are quite identical).

geometric pattern, Alhambra

stylised plant form pattern, Marrakesh

SHADOW AND REFLECTION
ablution fountain, Süleymaniye Mosque, Istanbul

ablution fountain, Süleymaniye Mosque, Istanbul

Mosques are generally provided with places for the ablutions required before prayer. Above is a drawing of the interior of the ablution fountain at the centre of the courtyard of the Süleymaniye Mosque in Istanbul, designed by Mimar Sinan in the middle of the tenth century AH (sixteenth CE). Sunlight strikes through latticed openings, reflecting off the surface of the water inside and back onto the interior wall surfaces. The result is, like the installation on page 156, an interior of intricate geometric shadow patterns, some of which are distorted by rippling water.

A mixture of shadows and reflections (in this case off the surface of the water contained inside the fountain's cistern) can produced an intricate overlay of patterns.

COURTYARD SHADE
Koza Han, Bursa, Turkey

Courtyards are instruments for the provision of shade.

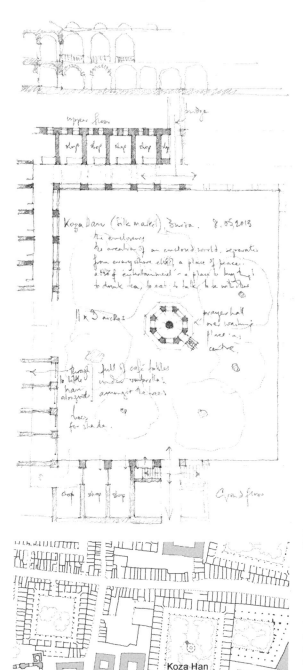

Koza Han is a silk market in the city of Bursa. It is in the form of a han or caravanserai, a courtyard type of market that has been used in the Middle East for centuries. The courtyard is lined with two storeys of shops accessed by shaded walkways. At the centre of the courtyard is a small mosque supported on pillars around its associated place for washing. Around the mosque are trees that provide the courtyard with dappled shade making it a comfortable and attractive place to sit and drink coffee or eat a meal. (On the left is the sketch plan I made in my notebook whilst sitting in the shade of those trees.)

Koza Han, Bursa

Koza Han is just one of a number of courtyard markets that make up the distinctive city plan of Bursa (left). Linking the hans are narrow souks – also shaded – lined with small shops.

Koza Han

COURTYARD SHADE
Alhambra, Granada, Spain

Court of the Lions, Alhambra, Granada, Spain

The courtyard, an open yet private space, is a useful aid to ventilation in hot countries. In Islamic houses and palaces courtyards are often supplemented with shady rooms or alcoves around their periphery (an echo of their Greek and Roman forebears; see page 59). In combination, a courtyard with its shady alcoves, and perhaps fabric awnings, can be the heart of a house, the place where it is most comfortable to be on hot airless days.

The layers of columns around the Alhambra's Court of the Lions (above) creates graded layers of shade, with some of the recesses in deep shadow.

A typical Islamic house (below) has a courtyard at its heart, open to the sky. Around that are shady alcoves for everyday life and entertaining.

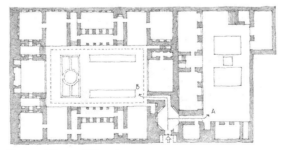

Monsoori House, Shiraz, Iran

DECORATING A ROOM WITH SHADOWS
Intersections, Anila Quayyum Agha

In 2014, the artist Anila Quayyum Agha exhibited an installation at the Grand Rapids Art Museum in Michigan. It was entitled 'Intersections', one of a series of similar shadow pieces.* The work consisted of an approximately 6' (2m) cube containing only a single source of light. The six wooden panels of the cube were fretted with patterns inspired by Islamic decorations from the Alhambra. The artist wrote of the work:

> 'The Intersections project takes the seminal experience of exclusion as a woman from a space of community and creativity such as a Mosque and translates the complex expressions of both wonder and exclusion that have been my experience while growing up in Pakistan.' **

Its architectural power lies in demonstrating that a room and its atmosphere can be decorated with no more than shadows. The cube projects a cubic room of shadows about itself, onto the floor and walls of the gallery.

The image illustrates the installation as exhibited in 2018. Light from the cube decorates the surrounding room with shadows of Islamic patterns.

* More of Anila Quayyum Agha's work can be found at: anilaagha.com/ (March, 2019).
** See: artprize.org/anila-quayyum-agha/2014/intersections (March, 2019).

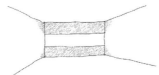

with horizontal windows

with tall narrow windows

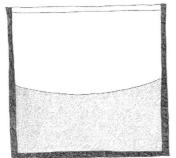

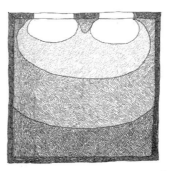

shadows reduced, more even lighting   room gloomy, corners dark

LE CORBUSIER
ARCHITECT OF SHADOWS

By way of a summary reviewing the wide variety of types and uses of shadows in architecture, all that is necessary is to consider just one twentieth-century architect whose work displays all of them. Le Corbusier is one of the most inventive and original architects of all time. There are many dimensions to his work. One of them is his use of shadows. He was the author of the quotation given at the start of this Notebook:

> 'Architecture is the masterly, correct and magnificent play of mass brought together in light. Our eyes are made to see forms in light; light and shade reveal these forms.'

In his work Le Corbusier considered not only the sculptural form of the elements of his buildings but also the qualities and types of the shadows they would cast. He used architectural masses to orchestrate shadows in the same way that a composer might aim to evoke light and shade in a piece of music.

It is one of the reassuring contradictions of Le Corbusier's work that while in the 1920s, in his 'Five Points for a New Architecture' (1926), he issued something of a vendetta against shadows with his promotion of horizontal windows (above), in his later career he became one of the most inventive architects to use them.

A ROMAN WORK OF SHADE AND LIGHT
Serapeum, Hadrian's Villa, Tivoli

One of the purposes of all the books on architecture that I have produced is to encourage architects (especially students) to become aware of the amazing potential of architecture by assimilating the work of other architects – from ancient times to the present – and to experiment in their own work with what they encounter.

Le Corbusier is one of the pre-eminent examples of the benefits of this approach to learning how to do architecture. One of the seminal periods of his life was when he went on journeys around Europe and to Turkey in the years 1910–11. During one of these trips he visited, and was fascinated by, Hadrian's extensive villa at Tivoli outside Rome. He studied the Serapeum (on this page and opposite) in particular detail.

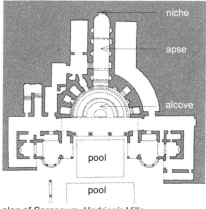

plan of Serapeum, Hadrian's Villa

'la lumière solaire est au fond de la caverne'
('the sunlight is in the depths of the cave')

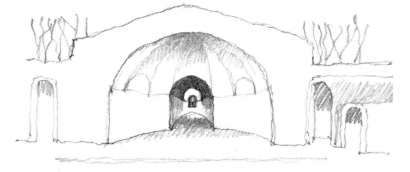

sketch 1, front alcove of Serapeum with view into apse, lit at its end by the sky

'ici la lumière solaire'
('here is the sunlight')

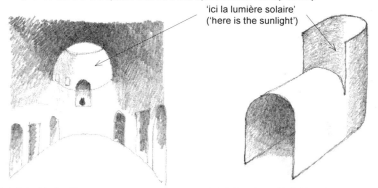

sketch 2, inside the Serapeum apse   sketch 3, schematic light strategy

A walk-through computer model of a reconstructed Serapeum is available at:
sketchfab.com/3d-models/serapeum-3127c970265e490da617c788d5863888 (December 2019).

'ICI LA LUMIÉRE SOLAIRE'
light enhanced by shadow

front alcove of Serapeum

speculative reconstruction

Serapeum apse with niche at the lit end

The Serapeum was one of the buildings of Hadrian's Villa that excited Le Corbusier most. It is a temple to an Egyptian deity and stands at the end of the Canopus – a long pool representing the River Nile. This temple has a half-domed open alcove fronted by a columned screen (top left and right). From this leads a long vaulted apse receding deep into the artificial hill against which the temple sits (above). You can see the alcove and apse, with its niche for the god's statue, in the plan (opposite top).

What impressed Le Corbusier most was the way in which the temple acted as an instrument for the manipulation of light and shade. He made numerous sketches in his notebook, three of which I have re-sketched on the opposite page.

The notes Le Corbusier added to his sketches indicate his interest in this building. Annotating the first sketch (sketch 1 opposite) he wrote, 'cà vaut l'effet de lumière' ('it's worth the light effect'). And under a similar sketch he wrote, 'cette lumière au fond de ce mont' (this light in the depths of the hill'). When Le Corbusier redrew this sketch later (he included it as part of his explanation of the 1948 scheme for La Sainte-Beaume in the *Œuvre complète*) he noted 'la lumière solaire est au fond de la caverne' ('the sunlight is in the depths of the cave'). In connection with both the lower sketches (sketches 2 and 3 opposite) he wrote, 'virtuellement c'est cette forme, et cet appel de lumière est beau' ('truly this form, and this admission of light is beautiful').

In this instance, Le Corbusier recognised that architecture could orchestrate shadow to enhance light. The key element here was the partial vault over the niche, which created a shadow frame for the shadow gradient at the end. As someone keen to learn the workings of architecture, it is clear that he stored this device in his memory for future use.

INSTRUMENT FOR PLAYING SHADOW
Ronchamp Chapel, Le Corbusier

Like an ancient burial chamber (above) the interior of the Ronchamp Chapel is contained shadow broken by light entering through gaps in the fabric.

The Chapel at Ronchamp (1954; in the east of France) illustrates some of the influence Le Corbusier drew from the Serapeum in Hadrian's Villa.

 The Chapel is the architectural equivalent of a musical instrument. Rather than playing sound (though during a service it does that too) the building plays light and various pitches of shadow. Light is admitted in various different ways into its dark interior. With its shadow gradients and thresholds, its framed shadows and shadow frames, and with its drawing with shadows, it is an object lesson in how shadows can contribute to the aesthetic experience and the poetry of architecture.

Inspired by the Serapeum, Le Corbusier uses light from above in the small side chapels, producing shadow gradients that dematerialise the rough surfaces.

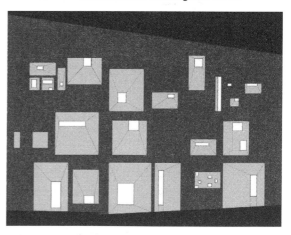

The celebrated south wall of the chapel is like a musical composition, with the ingredient 'notes' consisting of points of light framed by degrees of shadow representing degrees of silence.

FRAMED SHADOW FRAME
and shadow framed shadow

In the wall above the altar of the Chapel, there is a small rectangular 'window' containing a statue of the Madonna with Child (right; it is in the drawing of the interior opposite too). The statue stands inside a glazed concrete box, and can be turned either to face the interior of the church or, when an external service is being held on the eastern side of the chapel, towards the outside. From the interior the statue is in shade, framed by light but within a square shadow frame. From the outside its character is reversed: the statue stands in light but within a shadow frame of the darkness within the chapel.

Le Corbusier was not explicit about what he intended by presenting the Madonna with Child in two opposite states of light and shadow. Perhaps it was a reference to the Yin and Yang nature of belief, i.e. that everything consists of contrary opposites. Whatever his intent, the architectural orchestration of shadow was central to his poetry and his philosophy.

From the outside the statue of the Madonna with Child is seen in light framed by shadow.

While from the inside the statue is shadowed and framed by light. The square box within which the statue stands is itself framed by the dark shadow of the chapel's interior. It seems that Le Corbusier wanted to present this iconic image in two contrasting ways: shadowed against light; and lit against shadow.

DRAWING WITH SHADOW
Le Corbusier signing his buildings

Le Corbusier's buildings sometimes have images cast into their concrete. Where they occur, it is if Le Corbusier has stamped the building with his identity, his seal of possession, like noblemen used to stamp their seal into soft wax with their signet ring to acknowledge their authorship of a letter or legal document.

The above example is on the eastern (sunrise facing) side of the Unité d'Habitation in Marseille (1952). In the sunlight it draws the image of a man with shadows. The messages of Le Corbusier's shadow drawing are: 1. that architecture is the art that frames human life; and 2. that the dimensions of architecture should derive from human scale. Drawing the man in relief also tacitly makes the point that humanity is brought to life by the sun. Without it he would not be there.

Le Corbusier 'signed' some of his buildings with the image of generic man cast directly into the concrete walls. He often also included diagrams of his dimensioning system known as the Modulor.*

* See page 31 of the *Metaphor* Notebook.

PROJECTED SHADOW
La Tourette

On page 148 of the *Metaphor* Notebook I mentioned the glass screen walls designed by Iannis Xenakis for Le Corbusier's monastery of La Tourette as an example of the music metaphor in architecture.

Those glass screen walls with their concrete mullions and slim transoms also decorate the monastery's spaces with dynamic shadow patterns as the southern France sun moves slowly across the sky. They introduce not only the idea of music but its time too.

La Tourette, 1961

The glass screen walls of La Tourette (above), with their rhythmic 'undulating' patterns are an attempt to make architectural music. They were designed with Le Corbusier by a musical composer, Iannis Xenakis.

The irregularly spaced mullions also provide interior spaces with a play of shadows that constantly changes through the monastic day (below).

shadow play in one of the corridors

SHADOW CONTAINER
Zurich and Chandigarh

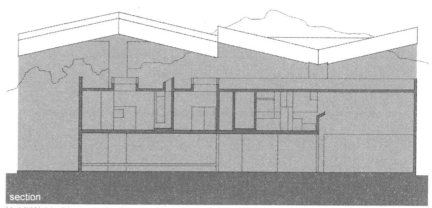

Heidi Weber Museum (Pavillon Le Corbusier), Zurich, 1960

In his design for the Heidi Weber Museum in Zurich, Le Corbusier demonstrates that the first step in conceiving a work of architecture can be the creation of a shadow container. Its geometrically complex canopies shelter and shade the museum below.

  The massive free-standing concrete portico of the Assembly Building in Chandigarh (below) is a gargantuan shadow container too, offering refuge from the searing Indian sun. (One of the other buildings Le Corbusier built in Chandigarh is a Tower of Shadows.*)

Above is the section given in Le Corbusier – Œuvre complète, Volume 7, 1957–65, 1965. (It is not exactly the same as was built.)

* See *Œuvre complète, Volume 8, 1965–69*, 1970, pp. 74–5.

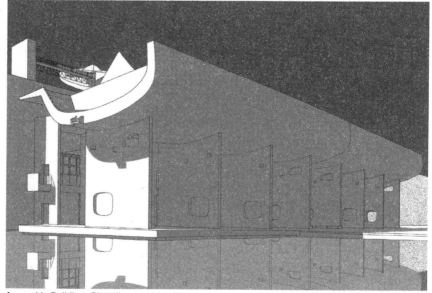

Assembly Building, Chandigarh, 1951–62

PROBLEMATIC SHADOW
the dead undercroft

In some instances Le Corbusier's buildings create shadow containers that are not so successful. One of his 'Five Points for a New Architecture' (1926) was that 'pilotis' (columns) could be used to lift a building off the ground. His argument was that, combined with a roof garden, this meant a building could, instead of subtracting from the ground's surface, double it. The natural ground could flow under the building while being repeated as a garden at a higher level on the roof. Unfortunately, being constantly in shade, vegetation will not grow in such undercrofts, and without the vitality of inhabitation (i.e. being used as verandas or entrance porches) they end up being used for parking and for rubbish bins. Below is the undercroft of the Marseille Unité d'Habitation.

In his 'Five Points for a New Architecture', Le Corbusier sketched the idea that, using pilotis, a building could have gardens both flowing under it and on its roof. The reality (below) is often different.

the undercroft of the Unité d'Habitation, Marseille, 1952

A SHADY PLACE TO SIT
Villa le lac, Vevey

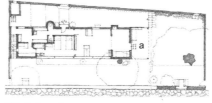

Le Corbusier built the Villa le lac for his parents in the early 1920s. It stands on the banks of Lac Léman in Switzerland. Its plan is a modern interpretation of an ancient Greek megaron (below left).

Villa le lac

Villa le lac, plan

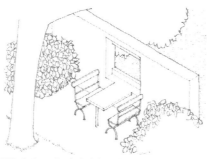

megaron, Tiryns, Greece

As with the megaron, Le Corbusier gave the villa a porch for sitting in the shade ('a' in the plan, left). Le Corbusier also provided an external dining space (below) shaded by a large tree and a stone wall, in which there is a glassless window framing (a shadow frame) a view across the lake towards the Mont Blanc massif.

The villa's external dining table (left) is shaded by a wall in which a window creates a shadow frame for the view of the dramatic landscape across the water (above).

See also *Villa Le Lac*, 2014; ebook available from Apple Books.

Villa le lac, shaded sitting place

PAINTING WITH SHADOW
expressed in Le Corbusier's choice of photographs

The volumes of Le Corbusier's *Œuvre complète* are illustrated with high contrast, carefully composed photographs. I have re-drawn some of them on this page, reducing their tonal range to black and grey. They show how Le Corbusier – a painter as well as an architect – enjoyed the ways in which he could use architecture to 'paint' compositions with light and shadow. The buildings shown are from his work in India.

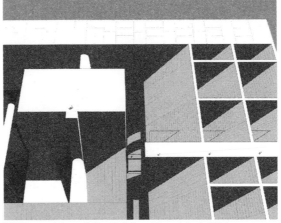

Mill Owners' Association Building, Ahmedabad, 1954

brise-soleil

Le Corbusier used his idea of the brise-soleil in many of his buildings in warm and hot climates (page 170). Recessing windows in deep alcoves means that warming from the sun is reduced. But the photographs Le Corbusier chose to illustrate them indicate that he enjoyed the visual patterns they made too.

The entrance of the Mill Owners' Association Building in Ahmedabad was illustrated in the pages of Le Corbusier's Œuvre complète as an abstract 'painting' in light and shadow (right).

Mill Owners' Association Building, entrance

SHADED ROOF TERRACE
Shodan House, Ahmedabad

The climate of India is at least as hot as that of Crete. When, in the 1950s, Le Corbusier designed the Shodan House in Ahmedabad (above) he gave it a roof terrace shaded by a concrete roof reminiscent of that ancient Minoan precedent illustrated earlier in this Notebook (page 78).

Note also that the house is provided with brise-soleil (see pages 167 and 170) to help prevent excessive solar heat gain inside its rooms.

With its orthogonal composition of concrete planes controlling sun and heat, the house is, to quote Le Corbusier himself: 'a masterly, correct and magnificent play of masses brought together in light'.

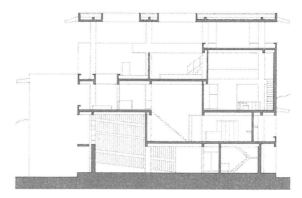

ANALYSING ARCHITECTURE NOTEBOOKS

LIVING IN A FAVOURED WORLD
Villa Savoye, Poissy

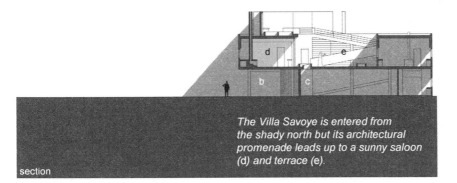

The Villa Savoye is entered from the shady north but its architectural promenade leads up to a sunny saloon (d) and terrace (e).

section

Creating an architectural sequence leading from shade into sunshine was a device that was often used in eighteenth-century neoclassical houses such as the House of Dun (see pages 102–4). But it can be encountered in modern houses too, such as the Villa Savoye in Poissy near Paris, designed by Le Corbusier in 1929.

In the Villa Savoye the architectural sequence of spaces is comparable to that in the House of Dun. It progresses from: the approach (a), also from the south; under the shade of the overhanging first floor (b), which acts as a porch; through the door into the hallway (c); up the ramp to the sunny saloon (d); and out to the first floor open courtyard (e). The sequence continues up a further ramp to the solarium on the roof (f, below).

Le Corbusier's strategy is a clever way of allowing the saloon to enjoy both sunshine from the south and the best views towards the north (towards the left in the plans below).

first floor plan (piano nobile)

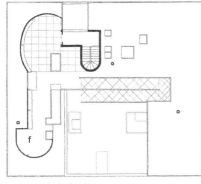

roof plan (solarium)

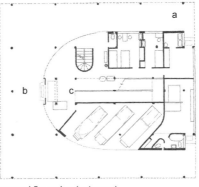

ground floor plan (entrance)

BRISE-SOLEIL
Le Corbusier breaks the sun

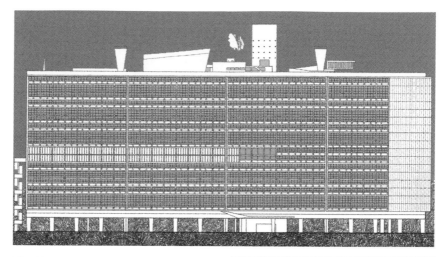

The Marseille Unité d'Habitation was completed in 1952. It was one of a number of apartment blocks planned by Le Corbusier to be built across France. The block was oriented north–south. As can be seen in the section below, apartments overlapped, with each having a view to both east and west. Though the apartments have large windows, Le Corbusier had a strategy for reducing the chance of overheating from the sun. He inscribed that strategy in a block of concrete near the building (right). It involved recessing the windows so that the sun would be 'broken' by overhanging balconies and horizontal slabs.

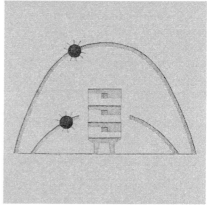

When the sun is low in the sky (right), during the winter and in the early mornings and evenings, the sun reaches deep into the apartments.

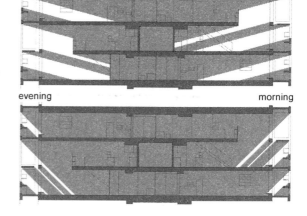

evening   morning

But when the sun is higher and hotter (right), the overhangs (brise-soleil) cut out most of its light and thereby reduce overheating.

SEARCHLIGHT
abbey church of La Tourette

The abbey church of the monastery of La Tourette near Lyons in France is a great rectangular container of shadow. Le Corbusier chose very carefully where that shadow would be broken by light. The most powerful instance is the beam of sunlight admitted through a rectangular opening in the roof.

The beam of sunlight is like a searchlight from heaven sweeping through the darkness occupied by the monks during their services.

SHADOW AND TIME (AND POETRY)
Hélène Binet – 'black square on the balcony'

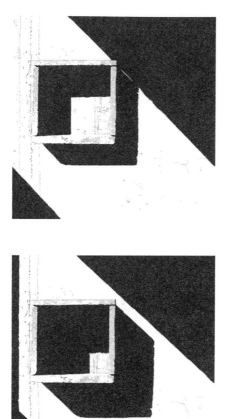

For a 2002 exhibition at the Deutsches Architekturmuseum in Frankfurt the photographer Hélène Binet decided to contribute a photo-essay on shadows in Le Corbusier's La Tourette monastery. For the book accompanying the exhibition she contributed an essay* in which she hints at the reasons for the choice of her examples. One of them she gave the subtitle 'Black square on the balcony':

> *'Private life in a cell. Light never comes into the room, but only onto the balcony that is situated at the end of the room. In that balcony space, on the wall that catches shadows, there is a small, square concrete box that is the only ornament in the balcony or the room. It brings you into a dialogue with the shadows. It breeds a strong sense of time, and time here is no longer ambiguous. The shadows are very sharp, they become a body. The shadow is a shadow of the box but also of the opening of the balcony; it is a kind of closure; it emphasizes the fact that you are very contained in that room.'*

On the left I have drawn three of Binet's sequence of seven photographs illustrating the development of shadows associated with that concrete box through part of the day.

It may be that Binet ascribes to this concrete box a poetry that was not consciously intended by Le Corbusier who saw it primarily as a practical place for a lamp, a book, a cup... But that is not the point. The poetry of shadows is where we find it. As architects the poetry of shadow is never quite under our complete control.

* Hélène Binet – 'Photographing Shadows at La Tourette', in Ulrike Brandi, Christoph Geissmar-Brandi and others, eds. – *The Secret of the Shadow: Light and Shadow in Architecture*, 2002.

ENDNOTE

To design a work of architecture that takes people through a sequence of different experiences of shadow is like composing a piece of music, except that the medium is not sound but the occlusion and modulation of light. Various devices can be used to create and manage the different kinds of shadow illustrated in this Notebook: cast shadows; shadow containers; contained shadows; shadow frames; framed shadows; shadow gradients and the rest. Different types of shadow can be layered like musical harmonies.

Not to take into account the contribution of this intangible, ungraspable, element of architecture would be to diminish the art. Shadows add aesthetic subtlety as well as reinforcing the perception of three dimensions. In low and high relief they are essential ingredients in the decoration and ornamentation of what would otherwise be flat surfaces. Shadows give surfaces texture. Shadows give rooms atmosphere, even drama. Shadows are powerful ingredients in orchestrating experience, and can elicit emotional responses: trepidation and relief; mystery and revelation; claustrophobia and release... Shadows contribute to architectural narratives.

Fixed lights cast fixed shadows. The shadows cast by the sun and the sky vary in intensity, direction and quality. When you hear a piece of recorded music it is always the same. When you hear that music live more than once, each time it is slightly different, with different performers interpreting it in subtly different ways, giving it their own light and shade. When you look at a painting or a piece of sculpture under modern museum lighting it is always the same. But architecture in daylight is never the same twice. It is like Heraclitus's river, which is a different river each time you step into it. Architecture changes constantly, inside and out. The sun moves; climates vary; in some, the weather changes minute by minute. (I once went to the Bodes Museum in Berlin. It was before the Wall came down. The galleries of Greek sculpture were side-lit only by the natural light from the windows. The effect was more affecting than with unchanging electric lights.)

Sciagraphy has always made architectural presentations more engaging. It was so in the time of the elaborate drawings of Beaux-Arts education (pages 35–42) and remains so with design assistive 3D computer software. But to exploit the full power of shadows in your work you need to consider their subtleties and how they will behave in the real world. You have to know how the sun will move across the sky and assess how it will change through the year. You have to understand the quality of light in the climate where you are designing. You have to judge whether, in that climate, shade is a cool (positive) or a gloomy (negative) place to be; whether shadow could be a useful accomplice to helping reduce a need for air-conditioning. You have to understand that shadows have different characters in different conditions: bright sunshine casts sharp shadows; overcast skies cast shadows too, fainter ones; reflected light creates shadow gradients... You should understand how rooflights will create different kinds of shadow to windows in walls; and be aware of the shadowy possibilities of translucent materials. You might even wonder about the emotional responses people might have to the shadow places or sequences of places you make... Neighbourliness suggests you should consider also how the shadows of your building might affect the lives of others around.

      There are many such subtleties to understand and consider. Few of them become apparent when you tick the 'shadows on' box in

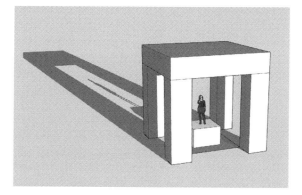

The shadow casting facility in some 3D software is useful but limited. There are many more subtleties to shadows than just those cast by computer-generated bright sunshine.

The actual quality of light and shade (as distinct from a 'visualisation') inside a prehistoric house*, as light percolates from a small opening above, is more accurately explored in a real model, under real light from the sky, than with computer 3D or graphics software.

* This my model of the interior of a house at Skara Brae on the Mainland of Orkney. See: Simon Unwin – *Skara Brae*, 2012.

popular software such as SketchUp (opposite) or Rhino. There, all you will see is a sharp cast shadow that moves with different times of the day and days of the year. To address the full potential of shadows you must consider your design in your imagination; and perhaps build real models and study them in real light from the sky (above). Then you will begin to appreciate the subtleties of shadow.

While I am putting together the material for one of these Notebooks, it always seems that the particular subject under scrutiny is the original and primary motivation driving architecture. This is clearly not true; but it is surely an indicator of the many powerful dimensions of this most rich and diverse creative endeavour.

It has not been particularly easy compiling this Notebook. Grasping at shadows, if you will forgive me, is notoriously futile! The photographer Hélène Binet (see page 172) wrote (with regard to her photography but it applies to my efforts in this Notebook too):

'Framing light is more natural. Shadows are something you have to play with. There is something about the shadow that always escapes you... Shadows, like breath, are almost ephemeral.'*

Nevertheless, I hope that I have managed to convey some of the many ways in which shadows play a part in the conception and experience of architecture. And that, in doing so, I have provided some prompts for your own experiments.

* See page 172.

ACKNOWLEDGEMENTS

I would like to thank: Jeff Balmer, Alanna
Donaldson, Fran Ford, Professor Wayne Forster,
Phil Henshaw, Chris James, Alwyn Jones,
Christopher Jones-Jenkins, Professor Stephen
Kite, Simon Lannon, David McLees, Lesley
McIntyre, Alan Paddison, Michael T. Swisher,
Rhian Thomas, Trudy Varcianna, Lili Wagner,
Dafydd Wiliam, Eurwyn Wiliam, as well as the
many students and colleagues who have over the
years, unwittingly, stimulated thoughts. And, as
always, my family...

BIBLIOGRAPHY

Aeschylus – *Agamemnon* (5thC BCE).

Anila Quayyum Agha – 'Intersections', available at:
artprize.org/anila-quayyum-agha/2014/intersections (March 2019);
see also: anilaagha.com/ (March 2019).

Gaston Bachelard, trans. Jolas (1964) – *The Poetics of Space* (1958), Beacon Press,
Boston, 1969.

Michael Baxandall – *Shadows and Enlightenment*, Yale U.P., New Haven CT, 1995.

Jérôme Bonin – 'Conarachne et Pelecinum: About Some Graeco-Roman Sundial
Types', *BSS* (British Sundial Society) *Bulletin*, Volume 27(i), March 2015,
available at: sundialsoc.org.uk/wp-content/uploads/Bonnin.pdf (September 2019).

Ulrike Brandi, Christoph Geissmar-Brandi and others, eds. – *Das Geheimnis des
Schattens: Licht und Schatten in der Architektur/The Secret of the Shadow: Light
and Shadow in Architecture*, Deutsches Architekturmuseum, Ernst Wasmuth,
Berlin, 2002.

Marcel Breuer (with Peter Blake) – *Sun and Shadow: The Philosophy of an Architect*,
Longmans, Green and Co, London, 1956.

Sir E.A. Wallis Budge – *The Literature of the Ancient Egyptians*, J.M. Dent & Sons,
London, 1914.

Norman F. Carver – *Form and Space of Japanese Architecture*, Shokokusha, Tokyo,
1955.

Joseph Conrad – *The Shadow-Line* (1916), available at:
gutenberg.org/files/451/451-h/451-h.htm (December 2019).

Caroline Constant – *The Woodland Cemetery: Toward a Spiritual Landscape*,
Byggförlaget, Stockholm, 1994.

Lawrence Durrell – *Spirit of Place: Letters and Essays on Travel*, Faber and Faber,
London, 1969.

Tim Edensor – *From Light to Dark: Daylight, Illumination and Gloom*, University of
Minnesota Press, Minneapolis MN, 2017.

T.S. Eliot – 'The Waste Land' (1922), available at:
poetryfoundation.org/poems/47311/the-waste-land (December 2018).

Arthur Evans – *The Palace of Minos* (1921), (in 7 volumes) Cambridge U.P., 2013.

Robin Evans – 'Mies van der Rohe's Paradoxical Symmetries', in *Translations
from Drawing to Building and Other Essays* (AA Documents 2), Architectural
Association, London, 1997.

Per Olaf Fjeld – *Sverre Fehn: The Pattern of Thoughts*, Monacelli Press, New York,
2009.

Per Olaf Fjeld – *Louis I. Kahn: Nordic Latitudes*, Arkansas U.P., 2019.

Robert Fludd – *Utriusque cosmi maioris scilicet et minoris metaphysica, physica atqve technica historia* (1617), available at: archive.org/details/utriusquecosmima01flud/page/n33 (August 2019).

E.M. Forster – *A Passage to India* (1924), Penguin, London, 1989.

Marie–Louise von Franz – *Shadow and Evil in Fairy Tales* (1974), Shambhala, Boulder CO, 1995.

Maxwell Fry and Jane Drew – *Tropical Architecture in the Dry and Humid Zones*, Robert E. Krieger, New York, 1964.

Sharon Gibbs – *Greek and Roman Sundials*, Yale U.P., New Haven, 1976.

John Hammond – *The Camera Obscura: A Chronicle*, CRC Press, Florida, 1981.

Robert Hannah and Giulio Magli – 'The Role of the Sun in the Pantheon's Design and Meaning', *Numen* 58:4, 1 Jan 2011.

Melissa Harrison – 'Nature Notebook', *The Times*, 16 November 2019.

Seamus Heaney – 'Alphabets', in *The Haw Lantern*, Faber, London, 1987.

Ernest Hemingway – *Death in the Afternoon* (1932), available at: la.utexas.edu/users/bump/images/Hemingway/death_in_the_afternoon.pdf (April 2019).

John Hollander – *The Substance of Shadow: A Darkening Trope in Poetic History*, University of Chicago Press, Chicago IL, 2016.

John M. Holmes – *Sciagraphy*, Pitman & Sons, London, 1952.

Homer, trans. Rieu – *The Odyssey* (c. 700 BCE), Penguin, London, 1946.

Tim Ingold – 'Commentary 1: On Light', in Papadopoulos and Moyes, eds. – *The Oxford Handbook of Light in Archaeology*, Oxford U.P., 2017.

James Joyce – *Ulysses* (1922), Penguin, Harmondsworth, 1968.

Franz Kafka, trans. Muir and Muir – *The Trial* (*Der Process*, 1914–15; 1925), 1937.

Frances Kéré, interviewed by Fiona Shipwright, in 'Of Clay and Community', *Mono. Kultur #46*, Berlin, Autumn 2018.

Stephen Kite – *Shadow-Makers: A Cultural History of Shadows in Architecture*, Bloomsbury, London, 2017.

The Koran, trans. J.M. Rodwell (1909), Phoenix Press, London, 1992.

Le Corbusier – *Œuvre complète, Volume 7, 1957–65* and *Volume 8, 1965–69*, Les Éditions d'Architecture, Zurich, 1965 and 1970.

Leonardo da Vinci, trans. MacCurdy – *The Notebooks of Leonardo da Vinci* (15thC), Reynal & Hitchcock, New York, 1939.

Leopardi, trans. various – *Zibaldone: The Notebooks of Giacomo Leopardi* (early 19thC), Penguin, London, 2012.

Federico García Lorca, trans. Merwin – 'The Song of the Barren Orange Tree' (early 1920s), in Francisco García Lorca and Donald M. Allen, eds. – *The Selected Poems of Federico García Lorca*, New Directions, New York, 1955, 2005.

Henry McGoodwin – *Architectural Shades and Shadows* (1904), available at: archive.org/details/architecturalsha00mcgouoft (March 2019).

Juhani Pallasmaa – *The Eyes of the Skin: Architecture and the Senses* (1996), John Wiley & Sons, Chichester, 2005.

Costas Papadopoulos and Holly Moyes, eds. – *The Oxford Handbook of Light in Archaeology*, Oxford U.P., 2017, available at: oxfordhandbooks.com/ (September 2019).

P. Planat – *Manuel de perspective et tracé des ombres*, Librairie de la Construction Moderne, Paris, 1899.

Pliny the Elder, trans. John Bostock and H.T. Riley – *The Natural History* (77–9 CE), Henry G. Bohn, London, 1855, available at: perseus.tufts.edu/hopper/text?doc=Perseus:abo:phi,0978,001:35 (January 2019).

Henry Plummer – *Light in Japanese Architecture*, A+U (Architecture and Urbanism), Tokyo, June 1995 (Extra Edition).

Raúl Rispa, ed. – *Barragán: the Complete Works*, Thames and Hudson, London, 1996.

John Ruskin – Description of St Mark's from *The Stones of Venice* (1853), available at: bartleby.com/library/prose/4427.html (November 2019).

Göran Schildt – *Alvar Aalto in His Own Words*, Rizzoli, New York, 1998.

Jeff Shannon, ed. – *Shadow Patterns: Reflections on Fay Jones and his Architecture*, University of Arkansas Press, Fayetteville, 2017.

William Chapman Sharpe – *Grasping Shadows*, Oxford U.P., Oxford, 2017.

John Soane, ed. Arthur T. Bolton – *Lectures on Architecture, as Delivered to the Students of the Royal Academy from 1809 to 1836 in Two Courses of Six Lectures Each*, Sir John Soane's Museum, London, 1929.

Ettore Sottsass, ed. Barbara Radice – *Design Metaphors*, Rizzoli, New York, 1988.

Tsuyoshi Tamura – *Art of the Landscape Garden in Japan*, Kokusai Bunka Shinkokai, Tokyo, 1936.

Junichiro Tanizaki, trans. Harper and Seidensticker (1954) – *In Praise of Shadows* (1933), Vintage, London, 2001.

Alexandra Tyng – *Beginnings: Louis I. Kahn's Philosophy of Architecture*, Wiley, New York, 1984.

Roger Tyrrell – *Aalto, Utzon, Fehn: Three Paradigms of Phenomenological Architecture*, Routledge, Abingdon, 2018.

Joseph Unwin – *Materialism Refuted: In a Series of Observations on Time and Eternity; Space and Extension; Matter and Motion; Light and Darkness: Out of which Observations, a Conclusive Proof is Drawn, that Neither the Universe, nor any of its Materials, can have always Existed*, W. Ford, Sheffield, 1829.

Simon Unwin – *Analysing Architecture* (1997), 4th ed., Routledge, Abingdon, 2014.

Simon Unwin – *Children as Place-Makers* (*Analysing Architecture Notebooks* series), Routledge, Abingdon, 2019.

Simon Unwin – *Curve* (*Analysing Architecture Notebooks* series), Routledge, Abingdon, 2019.

Simon Unwin – *Doorway*, Routledge, Abingdon, 2007.

Simon Unwin – *Metaphor* (*Analysing Architecture Notebooks* series), Routledge, Abingdon, 2019.

Simon Unwin – *Skara Brae*, 2012, available from Apple Books.

Simon Unwin – *The Ten Most Influential Buildings in History: Architecture's Archetypes*, Routledge, Abingdon, 2017.

Simon Unwin – *Twenty-Five Buildings Every Architect Should Understand*, Routledge, Abingdon, 2015.

Simon Unwin – *Villa Le Lac*, 2014, available from Apple Books.

Richard Weston – *Utzon*, Edition Bløndal, Hellerup, 2002.

Ivor de Wolfe (Hubert de Cronin Hastings) – *The Italian Townscape*, Architectural Press, London, 1963.

Frank Lloyd Wright – 'The New Architecture: Principles' (1957), in Edgar Kaufmann and Ben Raeburn, eds. – *Frank Lloyd Wright: Writings and Buildings*, Meridian, New York, 1960.

INDEX

ANALYSING ARCHITECTURE NOTEBOOKS

ANALYSING ARCHITECTURE NOTEBOOKS

'Silence to Light
Light to Silence
The Threshold of Their crossing
is the Singularity
is Inspiration
(where the desire to express meets the possible)
is the Sanctuary of Art
is the Treasury of the Shadows
(Material casts shadows shadows belongs [sic] to light)'

Louis Kahn – Page from a notebook (1972), illustrated
in Per Olaf Fjeld – *Louis I. Kahn: the Nordic Latitudes*, 2019.